Thinking Thoughtfully
Informal Logical Fallacies

A Practical Guide to Critical Thinking

Walt Thielen
Maricopa Community Colleges

Kendall Hunt
publishing company

Kendall Hunt
publishing company

www.kendallhunt.com
Send all inquiries to:
4050 Westmark Drive
Dubuque, IA 52004-1840

ISBN 978-0-7575-7101-5

Printed in the United States of America
10 9 8 7 6 5 4 3 2 1

Dedication

Dedicated to—

my wife, Barb, for her love, commitment, and editorial assistance

my son, Jake, and my daughter, Tori, for allowing me to experiment with their minds

my parents, who instilled in me the value of hard work and persistence

my brothers and sisters, who provided me with rich memories and valuable insights

my teachers and professors, who inspired me to follow in their footsteps

my students, who served as the inspiration for this book

my colleagues, who gave me their friendship and support

my readers, who sacrificed their time to gain new insights

my publisher, who helped put this book in your hands

Contents

Preface

As the title implies, this is a handbook that helps one uncover common thinking errors. It is intended to serve as a textbook for college courses that emphasize critical thinking. However, it is also written for the casual reader or general public. Some professors may wish to use it as a supplement to any course in communication, philosophy, writing, reading, logic, persuasion, rhetoric, and debate. Students in any course that focuses on critical thinking should find this handbook invaluable.

This is not a book of formal deductive logic, the logic of mathematicians and computer scientists. Rather, it is a book of informal logical fallacies and flaws—the types that are found in our daily conversations and experiences.

The introductory chapter discusses the objective of the book, reviews the importance of critical thinking, and explains how the book was developed. The second chapter provides a brief overview of reasoning. The final chapter offers some warnings about critical thinking, yet offers hope that such thinking is essential.

Between the two introductory chapters and the final chapter lies the heart of the book. These chapters are organized around six major themes of thinking errors. The third chapter reviews eight errors that often lead people to produce incorrect generalizations. Chapter four uncovers how people often identify the wrong causes for events—nine errors are explained. In the next chapter, seven errors related to authority are uncovered. In Chapter six, seven fallacies show how words are used to manipulate and persuade people. Ten errors, in Chapter seven, explain how unproven assumptions are used as a basis for making questionable decisions. Chapter eight examines twelve ways in which irrelevant and emotional appeals are used to override logical thinking.

Students should find this book easy to read because it is relatively short with a basic informal writing style and numerous concrete examples. It assumes no prior knowledge of logic or reasoning. While it is virtually impossible to provide a comprehensive coverage of all the known fallacies, the ones addressed in this book are those that are especially relevant for the vast majority of individuals.

Acknowledgements

I want to thank my wife for listening endlessly to my monologues about fallacious reasoning over the last two decades. Also, her suggestions as this book was being written were invaluable. I also want to thank my son and daughter for giving me frequent feedback on the clarity of the explanations and examples.

I also want to thank Dr. Marilyn Cristiano, chair of the Communication and Humanities Department at Paradise Valley Community College (PVCC), for her enthusiastic support of this project. Bob Bendotti, former vice president of learning at PVCC, and Lisa Kula, a colleague, also helped me secure district support for writing this book. I want to acknowledge the assistance of Maureen Stoss, a statistics instructor, for ensuring the accuracy of the statistical fallacies in this book. I also want to thank those students who helped me locate and purge the mistakes from the initial manuscript. As a professional courtesy, I sent the initial manuscript to my college president, Dr. Paul Dale, who not only acknowledged receiving it, but also actually updated me on his progress as he read it from cover to cover. Finally, the staff at Kendall Hunt Publishing Company provided outstanding editorial assistance and ensured that the book was published in a timely manner.

Dr. Walt Thielen

About the Author

Born in Iowa in 1960, Walt Thielen is the fifth of twelve children. Like his brothers and sisters, he grew up working long hours on a large dairy farm.

He began his education in a nearby one-room rural schoolhouse that included kindergarten through eighth grade, one teacher, and a total of 17 students. After the schoolhouse closed in the mid-1960s, he continued his education in a nearby town.

He graduated college with a bachelor's degree in speech communication and a master's degree in communication research. His doctoral degree is in education, with a specialty in higher education.

Dr. Thielen has been a professor of communication for more than 25 years, having taught at a major university, a small, private four-year liberal arts college, and various community colleges. He has taught a wide variety of courses, including the introductory human communication course, interpersonal, public speaking, small group, and business communication.

For the past 20 years he has taught at Paradise Valley Community College, a college of about 8,000 students in Phoenix. This college is one of the Maricopa Colleges, a community college system that is the largest in the country and that serves more than 250,000 students each year. It is also the largest provider of higher education in the state of Arizona.

He and his family live in Phoenix, Arizona. Dr. Thielen and his wife have two adult children, a son who just graduated college, and a daughter who is a senior in high school.

Chapter 1

Introduction

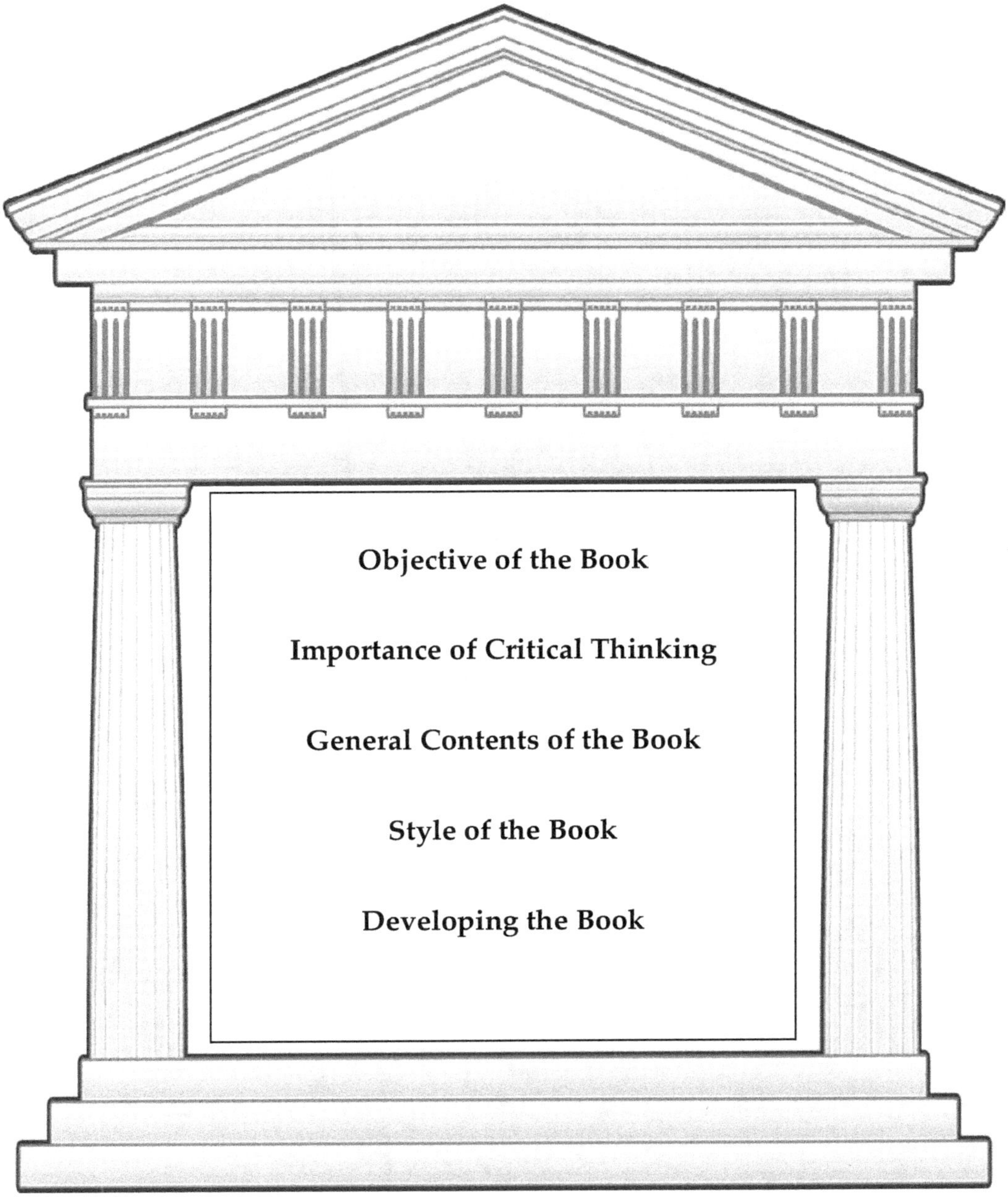

Chapter 1

Introduction

> *There's a mighty big difference between good sound reasons and reasons that sound good.*[1]
>
> —*Burton Hillis*—

The distinction that Burton Hillis makes between *good sound reasons* and *reasons that sound good* serves as the cornerstone of this book. This quotation summarizes what this book will attempt to accomplish—help individuals learn how to avoid poor reasoning. Poor reasoning can often initially sound good, but sound reasoning is not determined by eloquence. Instead, sound reasoning is that which uses accurate information and avoids fallacies, or thinking mistakes. This book is packed with mistakes, mistakes that commonly occur in reasoning. In this book, more than 50 common thinking mistakes, errors, or fallacies are pinpointed, exposed, and explained.

Although a fallacy will be explained in more detail later, a fallacy can be simply thought of as an argument or conclusion that sounds good, but that, upon more careful analysis, turns out to be unsound or defective. In short, a fallacy is a thinking error.

Objective of the Book

The objective of the book is to help teach people how to think for themselves—but to think in a way that avoids the common errors that afflict the masses of people. By pointing out how thinking errors and fallacies so frequently pollute common reasoning, we are less likely to commit them. And perhaps more importantly, we should become less likely to be influenced by others who use such fallacies. Ultimately, this knowledge should lead to more intelligent decisions.

While critical thinking affects decision-making, it also affects communication. The goal of this book is *not* to encourage people to increase the *quantity* of their communication; the goal is to increase the *quality* of their communication by first censoring their thoughts with tools of critical thinking. Once filtered, these thoughts are then ready to be expressed. This is not to suggest that all thoughts and feelings should go through a thorough internal censoring process. The goal of some communication is to develop social bonds or relieve emotional stress. In these cases, critical thinking is not necessarily of paramount importance. However, in many other cases, rationality and critical thinking is essential.

On occasion you may hear someone suggest that a particular person should study to earn a degree in communication because he or she already has the *gift of gab*. This is a simplistic and inaccurate perception of what the academic study of communication is all about. In fact, those people who supposedly have this *gift* are often the most incompetent communicators since they express their thoughts and ideas without first filtering them through reason and evidence. They might be eloquent and expressive, but accurate information and intellectual reasoning abilities are also requirements of effective communication.

To a large extent, effective communicators must first be effective thinkers. Becoming skilled in learning how to transfer to others one's ignorance, incompetence, or stupidity is worse than not communicating at all. There are many people who are expressive and passionate in communicating their thoughts and beliefs. And sometimes they are very effective in persuading others to accept their views. However, a significant percentage of these people are actually simply transferring their ignorance and faulty thoughts to others. In a sense, their minds are so infected with a disease of decay that they are unaware of their illness. Unknowingly, they infect others every time they speak.

Importance of Critical Thinking

Besides the general benefit of making more intelligent decisions and communicating more effectively, why is the study of faulty reasoning so important? There are at least four fundamental reasons why we should focus on learning how our reasoning becomes infected.

First, people naturally ask *why*. Why do we ask *why*? We ask *why* because it satisfies a powerful human need to know. In fact, you have a need to know whether you should continue reading this book. Without having reasons to read, you are unlikely to do so. One reason to continue reading is that this book deals with the most fundamental question ever posed, which is *why?*

Why is this important? This question is asked by children as young as two years old. Why does it get dark? Why do dogs bark? Why do birds fly? Adults ask physical force or intimidation, and people have little control over how they are governed.

However, in some cases people in democratic societies, were convinced, through propaganda and irrational arguments, to give up their power. For example, in Nazi Germany, Hitler actually obtained much of his initial power by his persuasive arguments, not through force or intimidation. Hitler's slaughter of millions resulted largely from a failure of critical thinking. His persuasion was powerful and emotional, but it was illogical and irrational. In many ways, millions of people were put to death because key people did not exercise critical thinking in the early days of his regime. In fact, the quote below embodies what he thought about the masses of people.

> *What luck for rulers that men do not think.*[2]
>
> —*Adolf Hitler*—

In summary, the study of critical thinking is important because it shows how people can satisfy their natural curiosity to find correct answers. Such study also helps to counter the natural tendency to let emotions overpower intellectual decision-making. Critical thinking also provides practical answers to practical questions. Studying critical thinking encourages one to become a better writer, reader, speaker, listener, and thinker. Developing these skills will partially immunize one from the damaging effects of deceptive political or advertising schemes and gimmicks. Investing more skillfully, selecting medical treatments more logically, purchasing products more carefully, and a host of other real-world abilities are just a few of the many practical reasons for developing critical thinking skills. Finally, studying and applying critical thinking helps to maintain a democratic and humane world culture.

General Contents of the Book

Early in this chapter, this book suggested that people not only have a desire to ask why, but really have an obligation to ask why. Asking why e-mail from an unknown source was sent to you, for example, is essential. Is the source of the e-mail attempting to divert you to another Web site, promote some product, or infect your computer? What is the content of the e-mail or Web site? Similarly, one should ask about the general contents of a book and the process used in developing that content before accepting its recommendations. The following brief analogy will attempt to provide those answers.

When we order a meal at a fancy restaurant for a special occasion, we rarely inquire about the process used to prepare the meal—we are simply concerned with the end result, the dinner itself. The fancy dinner is not the result of an accident. Although we don't give it much thought, we know that the chef in the kitchen took great care to select both the right ingredients and recipes. Like the chef at the fancy restaurant, we must also take great care when we prepare our daily meals. The best ingredients are of little value, though, unless they are prepared and cooked according to good recipes. On special occasions, we don't mind paying someone else to select the ingredients and cook the meal, but on a daily basis we usually have to do our own cooking.

Similarly, we sometimes pay others to do our thinking, but on a daily basis we usually have to rely on our own thinking to make decisions. In our thinking, we must take care to obtain the best ingredients—data, evidence, and information—when we make a decision. The best information, though, is of little value unless the best recipes of reasoning are used to develop answers, solutions, and conclusions. Accurate information mixed with sound reasoning leads to wisdom. And wisdom is required to solve problems effectively and make decisions skillfully.

The wisdom of a gourmet chef is the result of years of study, practice, and experience. The wisdom of a great philosopher, mathematician, or scientist also develops over time. A cookbook does not allow someone to become a gourmet chef, but it can help someone become a competent cook in a relatively short time. In the same way, this book cannot create an Einstein overnight, but it can allow one to become a competent thinker in a relatively short time. At the very least, it can help one avoid making basic thinking errors that so many people make on a daily basis.

The recipes for thinking in this book are not just any recipes—they are the best recipes formulated from the best chefs around the world. Furthermore, these recipes have been rewritten, described, explained, and organized in a way that should make sense for college students. Although somewhat similar to a cookbook of thinking, this handbook really focuses on how not to think. It primarily focuses on thinking errors that so commonly spoil our thinking.

Style of the Book

In preparing this handbook, highly technical descriptions and scholarly views found in academic journals and many textbooks have been replaced with more original explanations and concrete examples. Also, the writing style is more casual, conversational, and personal. Where appropriate, the formal style of writing using the third person pronoun has been suspended in favor of using the first or second person pronouns (I, me, my, you, etc.). These adjustments depart from conventional, formal scholarly writing, but often make the messages more psychologically engaging.

As a reviewer of college textbooks, I am intimately aware of the writing style that is common in standard textbooks, a writing style that students frequently complain about. I am also aware that my doctoral dissertation, like most dissertations, which demanded a scholarly and formal writing style, would probably put most students to sleep quicker than a sleeping pill. Also, as one who has worked with more than 6,000 college students in the last 25 years, I am quite aware that students find many standard textbooks boring. Hopefully the content and writing style in this handbook will make the flavor of reading it a bit more appealing than a traditional textbook.

Developing the Book

This informal handbook, however, was developed in a way that is very similar to a textbook. Both are the result of a systematic and rational development process. Why even mention the process used in writing this book? Some books give the impression that they are based on solid evidence and credible sources; however, in fact; they are largely fiction or opinions dressed up to appear as facts. For example, thousands of books have been written on astrology, but astrology has no basis in fact.

The method of research could be compared to the process of mining. For example, the miners in the California gold rush of 1849 traveled great distances to reach the mountains that held the riches they dreamed about. The kernels of gold, the nuggets of wealth, however, did not lie on the surface in plain view. Tons of rock, ore, and earth from deep in the mountains had to be processed to extract just one ounce of gold.

Likewise, each chapter in this book depended upon venturing deep into various sources, to process piles and piles of books, articles, conversations, and experiences. Scholars will recognize that highly respected authorities and sources in the field are cited. But, this book is written at a level that is appropriate for the casual reader. Every attempt was made to maintain the scholarly accuracy and precision of the concepts, while using examples and explanations that most readers will comprehend. For those interested in further mining these piles of raw materials, more than 100 entries in the endnotes and bibliography will allow one to locate the scores of original sources that helped to validate my explanations and examples.

Like the miners who were primarily concerned about the end result of wealth—this book washes away all the dust, mud, and debris from the original piles of research and writings and exposes just the specks of wisdom. Those specks were then melted together to form this book. Ayn Rand, a famous novelist, said, *Wealth is the product of man's capacity to think.*[3] The intent of this handbook is to help you achieve exactly that.

Chapter 2

Orientation to Reasoning

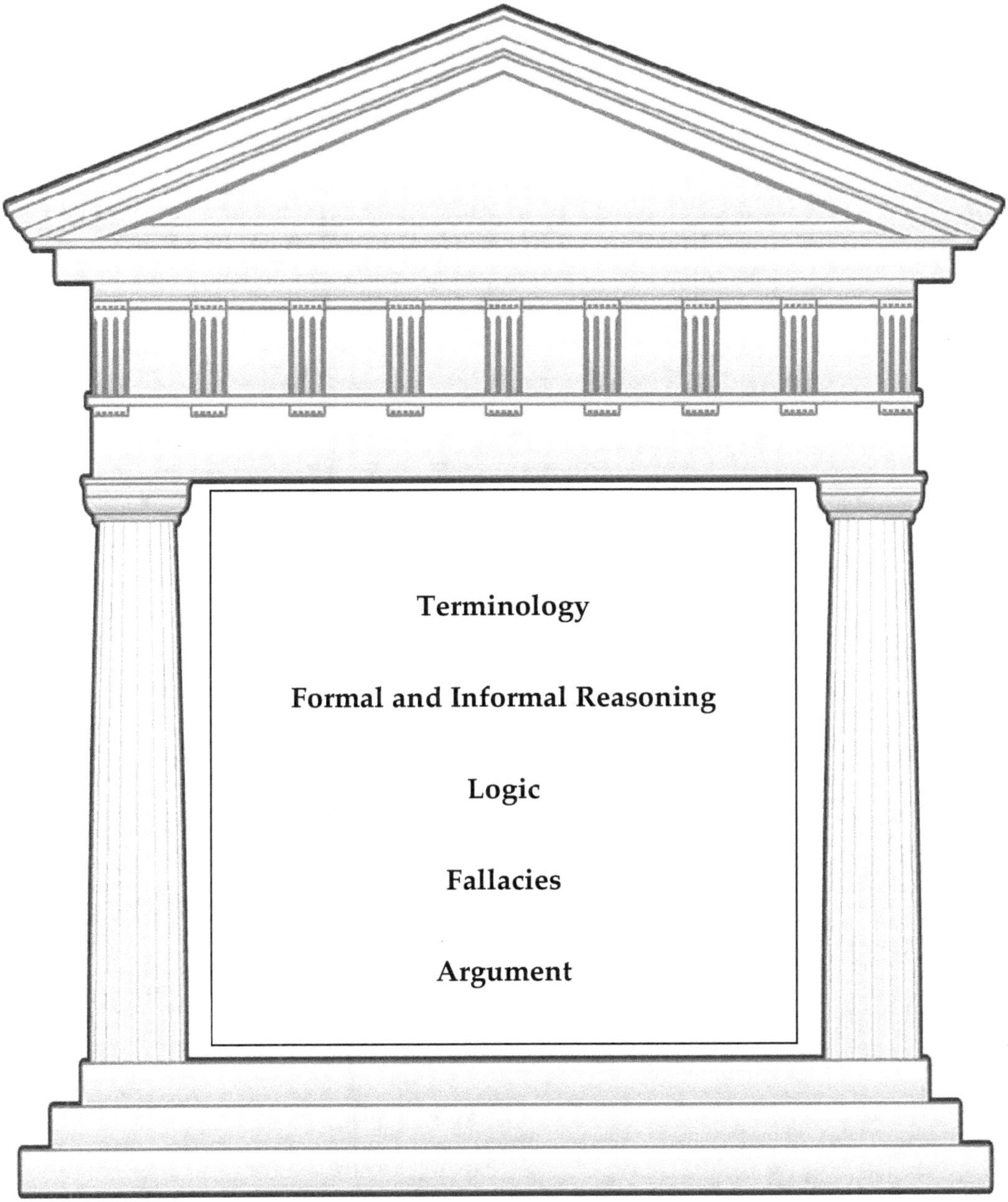

Chapter 2

Orientation to Reasoning

Before a surgeon and the nurses begin an operation, they have to undergo hours and hours of training on medical terminology. In fact, there are entire courses that focus almost exclusively on such terminology. *Doohickey, thingamajig,* or *whatchamacallit* are terms that lack the necessary precision. Although these terms may suffice in casual conversation, they are not adequate in the highly technical environment of a medical operating room. Similarly, when talking about reasoning, we must take care to ensure that terms are defined precisely so that genuine understanding occurs.

Terminology

Fallacious arguments are hiding everywhere. The first step in defeating our adversaries is to become aware of where they are hiding. Fallacies often mix in with the daily conversations of the crowd, making them difficult to expose. Once these enemies are found, they should be labeled as such so that others can readily identify them as well. In the case of fallacies, specific labels or terms will be given to these errors so that they can be more easily recognized. In a sense, attaching a term to a particular error not only allows one to talk with others about it, but it also makes it more obvious and visible to others.

For example, a casual observer of basketball often sees nothing more than players running up and down the court and shooting at random. A seasoned fan, however, sees much more; he sees dozens of specific actions and plays unfolding. When each of these plays is given a name and explained, the observer is more likely to actually pay attention and see these details as well. After learning about the plays, the viewer does not see a game of random shooting, but rather a game of picks, screens, give and goes, and a variety of other plays. Attaching a label to a basketball play or thinking error draws attention to itself. In a sense, giving a label to an item or idea is much like putting on a new pair of glasses. The new glasses, like the new words, allow us to see what we didn't notice before. This is why so many courses require that students become familiar with hundreds of terms. As such, each of the fallacies described in this book will be given a very specific label or term.

It is likely that you are already familiar with some of the errors in this book, but giving them labels will make them easier to recognize, talk about, and remember. In some cases, the terms used will be nearly self-explanatory. For example, a fallacy known as *either-or* simply reminds us to avoid forcing people to select only one of two options when more than two options exist.

In contrast, many of the fallacies will be more complicated and often have less-familiar names attached to them. In some cases, the fallacies have Latin names simply because they were coined nearly a thousand years ago in Europe. Many of these terms are commonly used today, both in classrooms and on the street. *Ad hominem*, *post hoc*, and *tu quoque* are just a few examples. In most cases, however, less archaic English terminology will be used. Learning the terms themselves is not the ultimate goal. The goal is to plant these terms, along with the explanations, in one's brain so that they are triggered by common examples found in political speeches, sales pitches, classroom lectures, news stories, or even daily conversations.

A few words on the front cover of this book, *Informal Logical Fallacies*, obviously indicate the major contents that lie between its covers. But, what do these words really mean?

To dress *informally* means to wear clothes that are casual or appropriate for everyday situations. We will accept this definition. That is, we will look primarily at reasoning that occurs in everyday normal conversations. We will not be concerned with a branch of reasoning that is known as *formal deductive logical reasoning*. This reasoning, though extremely important, will be left to the logicians, philosophers, mathematicians, and computer scientists. However, one way of learning what is considered informal dress is to compare it with formal attire. Similarly, learning what informal logic is can be understood by briefly comparing it with formal logic. What is the basic difference?

Formal and Informal Reasoning

Formal logic deals with deductive arguments, whereas informal logic deals with inductive arguments. For our purposes, only the most fundamental difference between the two needs to be understood. A deductive argument that uses valid reasoning and uses true premises will produce a ***TRUE*** conclusion.

True Premise: *A is less than B.*
True Premise: *B is less than C.*
True Conclusion: *Therefore A is less than C.*

In contrast, an inductive argument that uses valid reasoning and uses true premises will ***PROBABLY*** produce a true conclusion. Examples and explanations of inductive thinking are described on virtually every page in this book. But, the simplest way of understanding inductive reasoning is to remember it includes all types of reasoning that are not deductive. It includes reasoning that relies on examples, generalizations, analogies, statistics, assumptions and many other elements. Using simple statistics, such as the ones below, that infer some conclusion is one example of informal inductive reasoning.

Premise: There were 100 people at a bar.
Premise: 50 of these people were selected at random.
Premise: Of these 50 people, 70% were women.
Conclusion: Of all 100 people at the bar, about 70% are *probably* women.

A simple analogy to infer some conclusion is another example of using inductive reasoning. The example below is one that will produce a conclusion that is *probably* true.

Premise: My brother and I took the same class on critical thinking.
Premise: My brother and I earned the same grade in that critical thinking class.
Conclusion: It is *probably* true that my brother and I learned many of the same concepts.

Unlike the absolute statements that follow from deduction, inductive reasoning is simply concerned with how to increase the probability of making true statements. The majority of our daily reasoning deals with this inductive process. Police detectives, for example, employ inductive reasoning on a regular basis. That is, they look for specific clues to come to a general conclusion about who committed a particular crime. Put another way, they look at what is known (evidence, clues, etc.) and then make some prediction about what is unknown (who did it). With good evidence and valid reasoning, one's conclusion can have a high probability of being correct.

Logic

This distinction between deduction and induction is important, but it is also necessary to define the term *logic* with some care. The *Encarta World English Dictionary* defines logic as:

> *Sensible rational thought and argument . . .* [that] *aims to distinguish good from bad reasoning.*[1]

This is an acceptable definition; it will serve to limit what this book will explain. Applying this definition to actual situations is a bit more complicated. Someone may provide an argument, for example, and have it criticized as *not being logical.* Although it is true that some people let emotions and irrational thinking dominate their reasoning, two people can both use logical thinking and come to different conclusions. For example, consider the following question:

What is the logical choice to buy?

A. A 72-ounce bottle of soda that costs $2.99?
B. A six-pack of 12-ounce cans of soda that cost $3.99?

Most people would say the logical choice is the bottle of soda, because the cost per ounce is less than that of the cans. This is logical. Others might say that the cans are the logical choice because they are more convenient. This is also logical. So, before one can apply logic in this situation, logic needs to be defined as *cost* or *convenience.* In real-world situations, people often accuse each other of not being logical simply because they have no common definition of what exactly is being evaluated.

Similarly, because there is no common definition of what government is supposed to do, a book such as this cannot recommend that Democrats are better than Republicans or Conservatives are better than Liberals. Each voter ultimately needs to make those decisions based upon his values and individual circumstances. Valid reasoning simply allows one to make those decisions more carefully.

But, this book will investigate a number of political decisions that appear to involve questionable reasoning. Errors emanating from both major political parties will be uncovered. In order to use well-known examples, however, many will be drawn from the White House administration of the last eight years, from 2000 to 2008. The next edition of this book is likely to use examples from the next president, regardless of whether he or she is a Democrat or a Republican.

Those with strong partisan political views may find some of their conclusions challenged in one chapter but supported in another. Neither major political party is exempt from using questionable arguments. We must apply critical thinking equally to both our friends and adversaries.

In many other cases, however, what is *logical* is not subject to political values or one's individual circumstances. In the following example, logic produces one, and only one, correct answer. Despite the fact that many individuals frequently get this question wrong, logic provides us with one fail-safe correct answer. Read and answer the following question:

Suppose a man buys a horse for $70 and then immediately sells it for $80, and then immediately buys it back for $90, and then immediately sells it again for $100. How much did the man make on the transaction? $30? $20? $10? $0? Take a few minutes to provide your answer and then ask a friend the same question. Chances are that the two of you will not agree despite the fact that this is a simple math problem with one, and only one correct answer. Logic provides that answer. Anyone who does not select this answer is using reasoning that is not logical.

Fallacies

Now that *informal* and *logic* have been defined, let's look at another term on the front cover, *fallacies*. Fallacies are commonly defined as thinking or reasoning errors. A fallacy is much like pyrite—or fool's gold. Like gold, it is also shiny and attractive, but once inspected, it turns out to be worthless. Fallacies are like nuggets of fool's gold in that they are shiny, persuasive, and attractive; however, they turn out to be nothing more than skillful distractions or imitations of the real thing. The title of a classic book on logic is *Fallacy: The Counterfeit of Argument*.[2] Like the best counterfeit dollar bill or the best illusion a magician can create, a fallacy is an argument that tricks many in the crowd to accept an answer or conclusion that should not be accepted.

For this book, we are going to use a slightly broader definition of a fallacy. That is, a fallacy can occur when one's argument uses reasoning incorrectly OR reasons with incorrect evidence. That is, one could reason correctly, but start with evidence that is simply untrue. On the other hand, one could start with correct evidence and then reason incorrectly with that evidence. Although some scholars define a fallacy as being concerned only with the reasoning process, we will define a fallacy as any argument that uses either incorrect reasoning or false evidence.

The term *fallacy* comes from the Latin word *fallacia*, which means *to deceive*.[3] Many errors are the result of deception, but many others are the result of incompetence or carelessness.

Argument

Logic, *informal*, and *fallacies* are terms used frequently in this book. One last term requires clarification. The term *argument*, as used in reasoning, is not the type of debate or disagreement that many people think of when this word is mentioned. Instead, an argument is simply part of the reasoning and decision-making process. The following is an argument:

Premise: *Scientists tell us that if water exists, the possibility of life also exists.*

Premise: *Scientists have found strong evidence that there is water on Mars, the first planet they investigated.*

Premise: *Scientists have estimated that there are millions of planets.*

Conclusion: *Therefore, because there is a strong probability that water exists on other planets, I think there is a possibility that life exists beyond Earth.*

An argument is often defined as a conclusion that is produced from reasoning with premises or evidence. So, in our example, the first three statements are *premises*, and the final statement is the *conclusion*. Reasoning with the premises allows one to create the conclusion. So, when the word *argument* is used in this handbook, it involves reasoning with premises to reach a conclusion.

For the sake of clarity, the arguments and fallacies presented in this book will be quite obvious and pure. Extremely long arguments that are camouflaged with layers and layers of irrelevant details will be avoided. However, in many cases, real-world arguments are somewhat buried in cumbersome rhetoric, and irrelevant details have to be deleted to get to the core argument. The daily editorial page in the newspaper is a perfect example of this. Also in many cases, unstated assumptions have to be added to figure out how to even analyze the argument or editorial. When people are speaking, they often skip the obvious assumptions they are making in their arguments just as students will sometimes skip writing out the obvious steps they took to solve a math problem. To analyze arguments carefully, the listener might have to add those obvious or unstated assumptions. For example, consider the following bumper sticker.

Be Kind to Animals. Kiss a Hockey Player.

Obviously this bumper sticker is intended as a joke; however, it is a perfect example of making an argument, while leaving out some obvious assumptions. A logician, for example, would fill in those unstated assumptions or premises, known as *enthymemes,* with the following:

Premise:	*Be kind to animals.*
Premise (unstated):	*Hockey players are animals.*
Premise (unstated):	*A kiss is a sign of being kind.*
Conclusion:	*Therefore, kiss a hockey player.*

Although it is important to consider all the assumptions or unstated reasons in analyzing arguments, practically speaking, it would be ridiculous to talk this way. It would also be ridiculous to make a point of analyzing the details of our daily conversations with family and friends since the goal of most conversation is for social reasons. It would not be ridiculous, though, to carefully analyze the arguments and unstated assumptions of those who are obviously trying to persuade us, especially politicians, attorneys, journalists, and advertisers.

Chapter 3

Fallacies of Generalizations

Hasty Generalization

Insufficient Sample Size

Unrepresentative Sample

Spotlight

Whole Truth

False Analogy

Special Pleading

Sweeping Generalization

Chapter 3

Fallacies of Generalizations

Don't generalize! How many times have we been lectured on why we should avoid this terrible habit? But is it really such a terrible habit? There are certainly times when generalizing is inappropriate. For example, making broad sweeping generalizations, without qualifications (or without allowing exceptions) is often a sign of simplistic thinking. Similarly, not acknowledging that one is making a generalization is also a sign of verbal immaturity. In other words, one must follow a correct process in producing or deriving a generalization, but then must clearly acknowledge that the generalization is simply a generalization and not a statement of certainty. So generalizing incorrectly is a terrible habit, but when done correctly, generalizing is an effective and appropriate form of reasoning.

Unlike deductive arguments that produce conclusions of absolute certainty when valid reasoning and true premises are used, inductive reasoning, even with correct premises and valid reasoning, produces conclusions that have only a probability of being true. One of the most dominant forms of induction involves making generalizations.

But, all of the fallacies in this handbook are actually inductive ones. Poor induction can involve using information that is irrelevant or using assumptions that have not been proven as the basis for an argument. In fact, one chapter in this handbook is labeled *Fallacies of Relevance,* and another is *Fallacies of Presumption.* How is this chapter on generalizations different from other types of induction? *Fallacies of Generalizations* might use relevant information and even use assumptions that are well founded, but the error occurs in that the link between the premises is too weak to support the conclusion. That is, the reasoning itself is not very strong. Sometimes these errors are called *fallacies of faulty induction.*

There are many ways to describe the process of generalizing, but perhaps the key point is to remember that *known* information is used to make generalizations, conclusions, or predictions about *unknown* information. As mentioned previously, detective work, for example, involves generalizing in that evidence is gathered (known) and then a prediction is made as to who committed a particular crime (unknown).

Polling is another way of providing generalizations. The voting preferences of a few hundred people, for example, are gathered and studied. This is the known information. Then, by careful reasoning, one can predict with some degree of probability what the preferences of a larger group of people would be. This is the unknown information.

All of the fallacies in this chapter suffer from faulty generalizing. Sometimes a problem might be that the sample used to create a generalization was too small, or perhaps it was not selected correctly. Or, perhaps someone was making a generalization by using an analogy, but the analogy turned out to be comparing apples to oranges, rather than apples to apples. These are just a few ways faulty generalizations can occur. The focus of this chapter is to point out some of the most common errors we commit when we incorrectly generalize.

A popular quote by Mark Twain tells us: *All generalizations, including this one, are false.*[1] In other words, a generalization that says all generalizations are false is, in fact, false itself. Consequently, some generalizations are *generally* true. Avoiding the fallacies in this chapter will allow us to create these true types of generalizations.

Fallacy

of

Hasty Generalization

Last month I flew on American Airlines, and the plane was more than half an hour late. Today I flew American Airlines again, and sure enough, we sat on the runway for 20 minutes waiting to takeoff. I am never going to fly with them again. They are just so unreliable.

The *Fallacy of Hasty Generalization* is a flawed way of thinking, in which one makes a broad generalization based on limited experience or evidence.

As the name implies, people who are guilty of this fallacy are simply being too hasty, or too quick, in making a general conclusion or statement based on insufficient evidence. In other words, they are *jumping to conclusions*. They are not taking the time or care to ensure that they have adequate experiences to provide an accurate broad-based conclusion. Having two bad experiences on a particular airline, for instance, is hardly enough evidence for making the type of conclusion mentioned above. In fact, having *only* a 20-minute delay on these two flights might be better than the average airline.

People who commit this fallacy are assuming that all future encounters will be similar to the one or two they just experienced. Because they make this judgment prematurely, they shut themselves off from any future encounters. In a sense, they have reached a quick verdict and will not consider additional evidence.

This is probably one of the most common and dangerous flaws in our thinking. Unlike most scientists who carefully gather data in a systematic way and provide tentative conclusions, many people make snap decisions based on limited first impressions. As often quoted, *you don't have a second chance to make a good first impression*. This quote is unfortunately true because so many people commit the hasty generalization on a regular basis. That is, they use their first impressions to make major decisions or to develop strong opinions about a particular person. A major perceptual error involves placing too much stock in our first impressions. Frequently people etch a first impression of someone so deeply in their brains that it is difficult to ever remove that scar. Once this first impression is set, the mind works to gather information to reinforce this first impression, but ignores evidence that challenges it. The term *perceptual constancy* refers to a natural, but often damaging, habit of trying to keep our perceptions constant or stable, even in the face of new information. *Confirmation bias* is a term used to explain the human tendency to search only for information that confirms or reinforces previously held beliefs.

The tendency to make decisions too quickly or based on insufficient evidence occurs frequently. Many high school seniors, for example, will take a day or two to visit colleges or universities that they are interested in attending. In many cases, such seniors will visit a particular college and instantly notice the beautifully manicured campus, smiling faces from friendly students, and perhaps even the clear blue skies and a warm inviting breeze. Or, such seniors may visit a campus on a cloudy rainy day and see nothing but depressed looks from many students as they quickly run from class to class. Based on these very limited experiences, some seniors will make hasty decisions about what college to attend.

In fact, the reputation of the college, the qualifications of the faculty, the size of the classes, and other important academic variables are frequently overlooked. A high school senior who says he is attending a particular college because *everyone seemed so friendly* or *it just had a nice feel to it* is probably making a hasty generalization. This generalization, or decision, could easily involve $10,000 to $100,000. People often make important decisions without gathering sufficient evidence or taking sufficient time. Unfortunately, some people spend more time planning their vacations than planning their lives.

In contrast, most scientists use the scientific method in conducting their research. That is, they gather data in a systematic way to ensure the validity and reliability of their findings. The hallmark of the scientific method is all of the details of their experiments and conclusions are published so that other professionals can critique and replicate their findings. This process is self-correcting in that any errors or mistakes in the original research are found either by other professionals or in follow-up research.

Bypassing this process easily leads to hasty generalizations, and perhaps the most publicized scientific hasty generalization occurred in 1989, when Dr. B. Stanley Pons, professor of chemistry at the University of Utah, and his colleague, Dr. Martin Fleischmann, asserted that they had achieved nuclear fusion in a jar at room temperature. Such a discovery, if validated, would provide the world with an unlimited source of new and inexpensive energy. Unfortunately, Pons and Fleischmann were so anxious to announce their discovery to the public that the process of first publishing the results in scientific and academic journals was ignored. Such a momentous discovery turned out to be an embarrassment once other scientists found fundamental flaws in the research.[2]

The temptation to make hasty generalizations is easy to understand. The reason we make such quick decisions is because uncertainty is uncomfortable. Once we make a decision or conclusion, even if it is faulty, we feel comfortable again. In fact, this desire for comfort is one reason jurors often make up their minds so early in courtroom trials. Critical thinking, on the other hand, asks us to suspend our judgment and endure a period of uncertainty, and then slowly come to a conclusion once all of the evidence has been presented.

Of course, there are some situations in which decisions cannot be made slowly. Emergency room doctors, police officers, military commanders, and fire captains must often absorb as much information as possible and make quick decisions. In such cases, we would not suggest that these individuals are committing hasty generalizations.

In a court of law, committing a hasty generalization may lead to an improper verdict. Outside the courtroom, committing this fallacy often leads to prejudice or discrimination. One encounter with those of a particular race, religion, or culture is often used as the basis for making unwarranted assumptions about an entire group of people.

Again, the fallacy of the hasty generalization is simply making a decision based on limited evidence. Because hasty generalizations occur so frequently, one should examine exactly how these errors are made. But, before looking at these, it is important to remember that a hasty generalization can be thought of as a claim based on *insufficiency.*

One way this occurs is when we have an insufficient number of experiences to make a conclusion. In the case above, two flights is not adequate to make an accurate judgment. However, this fallacy is also committed if our experiences are somehow unusual, unrepresentative, or abnormal. For example, if the only time we fly is during the Thanksgiving holiday break, we are not experiencing the normal or typical conditions that occur during the rest of the year.

Because these two factors are so very common in hasty generalizations, we will actually label them as subsets or specific types of hasty generalizations. The *Fallacy of Insufficient Sample Size* and the *Fallacy of Unrepresentative Sample* are the two that occur most frequently. Whether it is making a decision about going to war, prematurely announcing a scientific discovery, or simply commenting about a particular airline, we should try to avoid making hasty generalizations.

Fallacy

Of

Insufficient Sample Size

I did a poll at my high school, and just 3 out of the 10 students I interviewed said they were planning to go to college. I thought it would be higher, but apparently only about 30% of the students at my school are going on to college.

The *Fallacy of Insufficient Sample Size* is a flawed way of thinking, in which one misuses statistics by making broad generalizations based on a sample that is too small.

As previously implied, this fallacy is actually one type of hasty generalization. This fallacy, though, employs statistical information. That is, both fallacies make generalizations based on insufficient evidence; however, this fallacy quantifies precisely the statements with the use of statistics.

In the example above, polling just ten students from a school that probably has several hundred or even thousands of students, is obviously insufficient. In our case, the number of students polled, ten, is known as the *sample*. The total number of students in the school is called the *population*. Common sense can easily detect such fallacies when the sample is clearly too small and the generalization is clearly too broad.

However, common sense can be used only as a rough guide. A basic understanding of statistics allows one to determine with more precision whether a fallacy is committed. For example, in the case of political polling, most people assume that one cannot make an accurate prediction of how 100 million people are going to vote unless at least 5%, or 5 million voters, are polled. Although this seems like common sense, it is entirely wrong. In some cases, less than one-tenth of one percent of the population could produce an accurate sample, whereas in some other cases, even 10% might not produce an accurate sample. Avoiding the fallacy of insufficient sample size is based on a number of variables.

For example, I can use a very small sample in some cases to accurately predict the composition of a larger body. Suppose my swimming pool holds 10,000 gallons of water, and I want to know the chlorine level of these 10,000 gallons. I can simply reach down and randomly remove one gallon of water, and test this gallon for chlorine. I can then predict with incredible accuracy the chlorine level of all 10,000 gallons based on the chlorine level in this one gallon. So, in this case, a sample size of just one one-hundredth of one percent would be sufficient.

Similarly, a doctor can accurately measure the cholesterol level of all ten pints of blood in one's body by taking a sample of an ounce. In this case, one ounce taken from ten pints (or 160 ounces) is more than adequate. Although it is far larger in proportion than the sample required for testing the pool water, it is still less than one percent of the total blood in one's body. In other cases, however, even a sample size of 50% is too small. An individual size bag of peanut M&Ms contains seven peanuts of various colors. I cannot, for example, look at the color of the first four peanuts (57%), the sample, and accurately predict the color of all seven peanuts with a high level of confidence.

Generally speaking, with a large population, a fairly small percentage of the entire population is of sufficient size for a good sample. If the population is small, then a larger percentage of the entire population is typically needed to create a good sample.

One can easily calculate how large the sample needs to be by consulting any basic textbook on statistics. Our concern, though, will not be with the exact calculations of various statistical measures themselves. Instead, our goal is to understand what they mean.

When polling results are reported, almost always a *margin of error* is included. Sometimes the *level of confidence* is also presented. Both of these terms have very precise meaning and refer to specific calculations.

When a poll announces, for example, that Candidate A is the favorite of 52% of the voters and that Candidate B is the favorite of 48% of the voters, the announcement is usually followed with a statement, *margin of error + or – 3%*. In addition, pollsters refer to the *level of confidence* in their poll. Often pollsters will use at least a 95% level of confidence. This means that the pollster is at least 95% certain that his prediction will fall within the stated margin of error. In other words, the pollster is 95% certain that candidate A will receive a low of 49% to a high of 55% of the total votes.

Comparing the final presidential polling predictions with the actual election results from national survey organizations over the past several decades has shown that their polling is accurate. That is, they were within the margin of error virtually every time. From 1956 to 2004, the average error was less than

3%.[3] On a rare occasion, a poll will be outside its margin of error. When this happens, enormous media attention is given to this so-called *mistake.* In fact, it is usually not a mistake. For example, a pollster, who uses a 95% level of confidence will admit, and even predict, that his results will be outside the margin of error 5% of the time. Put another way, reputable polls are correct a very high percentage of the time. This accuracy increases even more when the results of several polls are combined.

People often misread polls. A poll showing that Candidate A is at 52%, and that Candidate B is at 48% (assuming a margin of error of 3%), is NOT predicting that Candidate A will win. Unfortunately, many people misread the polls and often incorrectly conclude that the pollster was wrong if Candidate B wins.

Hasty generalizations often result from this fallacy of insufficient sample size; however, they also often result when the sample is not representative of the population, or a cross-section of the population is not used in making the prediction. This *unrepresentative sample fallacy* is the next one we will discuss.

Fallacy

of

Unrepresentative Sample

> *I just don't think doctors know what they are talking about. They say smoking causes cancer, but I don't think they really know. My neighbor is 90 years old, and he has smoked two packs of cigarettes a day since he was 20. He is as healthy as a horse.*

The *Fallacy of Unrepresentative Sample* is a flawed way of thinking, in which one looks at the characteristics of a sample that fails to contain a good cross-section of the population because it was not randomly selected, and then assumes that the characteristics of this sample can be generalized to the larger population.

The example above about the 90-year-old smoker actually uses a sample that is clearly too small, but more importantly, the sample is not representative of smokers in general. In other words, we cannot simply select a person who happens to be near (our neighbor) or someone we like (our friend).

In order to produce a valid generalization, we must base it on using enough samples and using samples that are representative. Likewise, if we are trying to refute the smoking-cancer generalization, we must cite enough samples and use samples that are also typical. There is a well-established claim that smoking is a major cause of cancer, and trying to deny this fact by citing a couple examples that are not typical (neighbor) does very little to weaken the claim. Generalizations are simply that—they acknowledge that exceptions usually exist.

Denying that smoking is not a significant cause of cancer by referring to an example that is not typical or representative of the population as a whole has virtually no merit. In fact, many people believe that they are genuinely refuting high-quality opinions or generalizations by simply producing one or two abnormal exceptions.

The previous *fallacy of insufficient sample size* dealt with a sample size that was too small, whereas this fallacy deals with a sample that is not well selected. One cannot, for example, survey the opinions of 10,000 basketball fans in one city, a very large sample, and assume they represent the views of all fans in the country. Obviously, the opinions of fans in one city would be highly biased and clearly not reflective of fans in general. In this case, the fallacy of insufficient sample size is not committed, because the sample is plenty large. However, no valid generalization about the fans' opinions can be made because the sample would not accurately represent the opinions of the fans in other cities. Like the previous fallacy, this is one of the more common errors that infect our thinking. In many cases, our conclusions suffer from both of these fallacies.

For example, one could reasonably determine the percentage of various nuts in a bag of assorted nuts of peanuts, almonds, and cashews (the population) by grabbing a handful of nuts (the sample) and then looking at the percentage of nuts in the sample. This could be done because the nuts in the bag are randomly distributed, and a simple handful is likely to capture the randomness of this mixing. On the other hand, suppose three individual bags of nuts are purchased, with one being all peanuts, another being all almonds, and another being all cashews. Emptying all the bags into the same jar one at a time would not produce an even distribution of nuts, but rather layers of nuts. In this case, grabbing a handful of nuts would result in having more nuts from the top layer and fewer from the bottom layer. In this case, simply grabbing a handful of nuts is not likely to reflect a cross-section of the population.

Similarly, suppose we have a rather odd cake, a cake that is shaped like a cube, one-foot square, with 24 layers of cake. The baker says that each layer will be of a different color or flavor, although one layer might be one inch thick whereas other layers may be as thin as one-quarter inch. Frosting covers the finished cake so one cannot see the layers that are hidden beneath. A cake that is 12" X 12" X 12" equals a total of 1,728 cubic inches. Despite the fact that the cake contains 1,728 cubic inches, it would be possible to predict the precise composition of the cake, in terms of color or flavor, by using a sample of less than one cubic inch, or less than one-tenth of one percent. All one would need to do is to take a standard twelve-inch clear drinking straw, insert it vertically into the cake, remove the straw, and then measure the thickness of each color. So an analysis of a sample that is extremely small relative to the entire population could produce a very accurate generalization about the entire cake. This would be similar to the example cited in the previous fallacy about using just one gallon of water to accurately reflect the chlorine level of 10,000 gallons in the swimming pool. One key, then, in making an accurate generalization is to be certain that the sample is representative of the larger population.

Now if the cake was not uniformly layered, but had a marbled consistency, a much larger sample would be required. In fact, in this instance, predicting the consistency of the entire cake might require a sample size of at least 10%, and the accuracy of that prediction might be lower than the results from using a sample of 1% of a uniformly layered cake.

In most cases one tries to ensure that the sample is reasonably randomized so that a reasonably accurate prediction about the population can be made. If the sample gathered is truly selected at random, a larger sample usually produces a conclusion with more accuracy, or a smaller margin of error.

Most of these principles apply to medical testing as well. Failing to obtain an adequate sample size commits a fallacy; however, failing to ensure that the sample to be tested really represents the population makes the generalization invalid as well. If a sample includes individuals who were selected at random from the general population (and assuming the sample size is of adequate size), then the generalization can be applied to the general population; however, it could not be generalized to a specific population, such as men older than 75 years old. In this case, the samples would also have to be selected at random from men older than 75.

One of the most common ways that the public is deceived occurs when the mass media, advertisers, or politicians present *evidence* and then suggests that this evidence shows what people are really thinking or wanting. For example, suppose a politician says that he received more than a 1,000 letters, and over 95% of those letters support the death penalty. If this politician then suggests that 95% of the public supports the death penalty, he has clearly made an unqualified generalization, because he is relying on a sample that is not a cross-section of the entire public. Almost always, only those who have extreme opinions take the time to write to their representatives. Those who have extreme opinions are far more likely to express how they feel than those who are neutral, undecided, mildly supportive, or mildly opposed. In most cases, these *middle-of-the-road* people represent the vast majority of the population. Typically, only the extremely dissatisfied and irate customers complain. Because only one percent of the consumers complain does not suggest that the remaining 99% are blissfully happy with a product or service.

The *hasty generalization* and *unrepresentative sample* fallacies are so closely related that it is not possible to discuss one without referring to the other. Generally speaking, a smaller sample size can be used if one knows with certainty than the sample is uniformly similar to the larger group or population. However, when this is not known, as in the marbled cake example, a much larger sample would be needed to ensure that a representative slice was actually obtained.

Is the sample size large enough? Is the sample truly representative? Critical thinkers ask these two questions of every conclusion they encounter; without proof that both these conditions are met, accepting the conclusion is dangerous.

Fallacy

of

Spotlight

A year ago there was a TV news story about a high school basketball player who just fell over and died after practice. Then last month I read in the paper about a football player who was only 16 and died of a heart attack at halftime. Our local news channel is even going to air a special report on these terrible tragedies. It is pretty clear that high school sports are becoming very dangerous.

The *Fallacy of Spotlight* is a flawed way of thinking, in which one makes an unqualified generalization by focusing too much on stories that are repeated and sensationalized by the media.

As explained previously, making generalizations from limited information is a common reasoning error. To improve the quality of the generalization, it is important to obtain a good sample or cross-section of the population. Failure to do this results in the fallacy of unrepresentative sample. This spotlight fallacy is actually just one type of unrepresentative sample. That is, it is not representative, because the sample includes only those examples that are interesting, or sensationalized by the media. These sensationalized snapshot examples are then publicized as representing the *big picture.*

The term *spotlight* is drawn from the media, in which a camera focuses upon the most sensational stories, rather than on the more common and mundane experiences of daily life. In many ways, the media presents examples that are exciting, abnormal, or unusual. While tabloid journalism is especially guilty of this, mainstream newspapers and TV networks also magnify such human-interest stories.

In the earlier example, excessive media coverage of a handful of deaths leads many people to vastly overestimate the risks of high school sports. A heart-wrenching story about the death of a handsome young man who was also the captain of the football team and an honor student is shown on the local news. Within a few days, the national news airs the same story, and within weeks, even family members and friends are asked to appear on afternoon talk shows.

All this attention on such unusual cases often leads the public to believe that the highly unusual is actually the norm.

For example, every school shooting is widely and repeatedly reported across the country. Usually these accounts are embellished with vivid or gruesome details and eyewitness testimony. Many will remember the Columbine High School massacre in 1999 when two high school boys went on a shooting rampage and killed 12 students and a teacher, and wounded 23 more.[4] Nor will many forget the worst school shooting in the United States in 2007, when a South Korean college student gunned down 32 of his classmates.[5]

Despite these tragedies, considering the sheer number of people concentrated on these campuses, schools and colleges remain some of the safest places in the country. In the United States, more than 75 million students attend thousands of schools each day.[6] Fewer than one-tenth of one percent of all murders committed in this country took place at a school.[7] Boring statistics like these, however, rarely get much media attention. In fact, there is much truth to the saying that *no news is good news*. In many ways, the focus of the news is on the abnormal or negative. A newspaper will obviously not report that 99.999% of students safely endured another school year.

Similarly, this spotlight fallacy is probably one reason why many people overestimate the murder rate and underestimate the suicide rate. Murders typically receive wide and repeated media coverage, whereas suicides are often kept out of the news. Consequently, many people do not realize that nearly twice as many suicides as murders are committed each year in the United States.

The spotlight fallacy often feeds mass hysteria. For example, millions of people receive vaccinations each year, and they therefore protect themselves from developing devastating diseases. This news is rarely reported with much fanfare. However, of those millions of people, there are sometimes a handful of people who experience serious side effects. These side effects often become front-page news. Some people therefore underestimate the massive benefits of vaccinations and overestimate the statistical probability of severe side effects.

Similarly, the public cannot escape the media bells and whistles when someone wins a jackpot. TV newscasters, journalists, and casino owners are always on hand to publicly show the happy gamblers receiving their million-dollar checks. However, the media gives no corresponding attention to the more common gambling activity that is occurring far more frequently. Millions of dollars are quietly lost everyday in casinos and in lotteries across the country.

The terrorist attacks of September 11, 2001 created images that are forever burned into most American's minds. The film clip of the airliners crashing into the two towers of the World Trade Center, and the ensuing collapse of those buildings has probably received more broadcasting time that any other in history. The simple fact that the media's account of the attack was visual and highly emotional guarantees that most Americans will never forget this attack. The repeated airings of these film clips ensured that most Americans saw the attack dozens, if not hundreds of times.

These images, though, nearly buried other important parts of the attack—the airliner that crashed into the Pentagon and the airliner that crashed in a Pennsylvania field. Obviously one would expect more coverage of the collapse of the towers simply because more people were killed there than in the other two attacks, but maintaining a reasoned perspective is also important. For example, although the attacks were horrendous acts of terrorism, along with the 3,000 people who perished in those towers, another 3,000 Americans died that day of heart disease and smoking-related causes. More than 3,000 continue to die from such causes EVERY day.[8,9,10]

This spotlight fallacy is related to several other fallacies of faulty generalizations. As discussed previously, the fallacy of the unrepresentative sample, for example, occurs when one reaches a faulty general conclusion by looking at a sample that doesn't reflect the whole. For example, reaching a conclusion about how Americans in general stand on various social issues just by sampling voters in a few conservative southern states would involve this fallacy.

This spotlight fallacy is just one way a sample becomes unrepresentative. The spotlight fallacy depends upon looking only at those samples that are spotlighted, or widely or frequently presented in the media.

This fallacy is also known as the *fallacy of misleading vividness*. As the name implies, vivid and dramatic examples presented by the media often lead people to form generalizations about the world that are wildly unrealistic and inaccurate.

Fallacy

of

Whole Truth

> *Dad, you should have been there! Coach put me in with three seconds left, and I shot the basketball from 30 feet. It went through the hoop at the buzzer, and we won the game!*

The *Fallacy of the Whole Truth* is a faulty way of thinking, in which one provides a claim, but omits relevant information that might influence whether the claim should be accepted.

In the example above, much of what the son says is true. It was true that he was put in the game at the very end and shot the ball from 30 feet. It was also true that his team won the game. However, what the son didn't tell his father was the reason he was put in the game—he was put in the game because his team was ahead by 20 points with three seconds left in the game. The son's claim that *we won the game* is technically true. His single basket, however, was not the reason his team won the game. With his three-point shot, his team won by 23 points. In other words, the whole truth was not provided because the son left out some very relevant information when he told the story to his father. The son obviously left the impression that his basket was the game-winning shot.

The whole truth might not be conveyed when a defendant appears before a judge. The defendant's claim that, *I was speeding because someone was chasing me,* might be true. However, if the reason for being chased is left out of the story, the judge may be left with the wrong impression. In this case, what the defendant failed to mention was that he was speeding, and the *someone* chasing him was actually a police officer.

Telling the judge that the reason for the accident was that *some guy pulled out right in front of my car* may be entirely true. But failing to mention that you were going through a red light at the time obviously commits this fallacy.

This fallacy is especially common in marketing and sales. Advertisers are quick to explain all the wonderful features or benefits of a product, but they rarely admit to any defects or disadvantages. Individuals, especially those in sales, commit this fallacy so frequently that the public actually expects it. Most assume that a used car dealer, for example, will intentionally and routinely employ this fallacy for deceptive reasons. Other times the fallacy is committed unintentionally by sales people because they honestly do not know the disadvantages of the products they are selling. In fact, some sales training intentionally withholds any negative information about their products so that the sales team is left to believe that their product is nearly flawless. So, the sales trainer might use the fallacy intentionally by deleting any negative information about the products during training. As a result, the sales team unintentionally commits the fallacy by being knowledgeable enough to give only partial information to the consumer.

This fallacy is also committed by various organizations that seek to promote a particular political philosophy. Proponents will often present only their side of the story. One of the most common ways politicians advance their cause is to avoid telling the whole story. Politicians, for example, might cite polls, evidence, or studies, which seem to support their position; however, those same politicians often omit or fail to mention many other studies that oppose their position. Equally deceptive is the failure of politicians to mention that much of the evidence supporting their positions might come from organizations associated with their campaigns. Political and religious organizations often give the impression that the studies supporting their positions are conducted by independent and unbiased agencies. Frequently, this is not the case.

Similarly, many talk-radio stations advocate extreme political agendas, yet give the impression that they present the news in an objective fashion. Consequently, many listeners of such programs believe the *truth* is being presented when, in fact, the information is highly biased and politically skewed. Most ardent listeners of talk-radio shows are rarely presented with the *whole truth.* And when they consume only biased information, they too often develop unwarranted biases. Obtaining the whole truth requires relying on a variety of credible news sources.

The fallacy of the whole truth is often committed because many people acknowledge that *errors of omission*—leaving out some relevant information—is less ethically troublesome than *errors of commission*—adding false or inaccurate information. Related to errors of omission is the use of equivocation, which involves being somewhat deceptive by being intentionally ambiguous or unclear. When people are faced with the choice of telling a lie or equivocating, most people will equivocate. In other words, when the husband is asked what he thinks of his wife's new dress, he is unlikely to say *it looks awful*. Even though that is what he might think, he will often choose to equivocate by saying *it certainly is colorful.*

Why do so many people intentionally use this fallacy, and why do so many people fall for its deception? Quite simply, this fallacy works. That is, a significant percentage of the buying population is either not sophisticated enough, not well informed, or simply too trusting to defend against this tactic. In sales, for example, avoiding telling the whole truth about a product or service is standard practice. Unsophisticated buyers often believe the sales pitch without being suspicious, or lack the ability to adequately research a product's flaws before buying.

This *fallacy of the whole truth* is sometimes known as the *fallacy of the half-truth*. *Cherry-picking* is another term used to describe the act of selecting, or picking, only that information that is ripe or helpful to one's cause. Of course, this also involves ignoring those cherries that are not ripe, or that information that is detrimental to one's cause.

The fallacy of the whole truth, like many other fallacies, attempts to fool people into accepting a claim or generalization by presenting only the evidence that supports the claim. Both a sufficient amount of evidence, and evidence that comes from both sides of the story must be considered before one can produce a reasonable generalization.

Fallacy

Of

False Analogy

My twin and I earned similar grades in elementary school. We also earned similar grades in junior high school. In high school our grades were almost identical. Even in college we both graduated with high honors. Given that we are alike in so many ways, it is likely that my twin and I will become pregnant at the same time.

The *Fallacy of False Analogy* is a faulty way of thinking, in which one compares two items or concepts and finds them similar in some ways, and then uses these similarities to incorrectly conclude that they are also similar in other, but irrelevant ways.

Reasoning by analogy is a very relevant and useful form of reasoning. In fact, it is one of the powerful arguments used in the courtroom. An attorney who shows that his case is similar to another previously decided case uses that precedent, or decision, in arguing that his client's case should receive the same verdict.

However, in order for analogous reasoning to be strong, one must be sure that the characteristics on which the items are similar, are relevant or predictive of other similarities. For example, in the first situation above, the academic performances of two individuals are compared at several levels, and they are shown to be similar. Given these similarities, it is appropriate to reason that these two individuals are probably going to be similar in other *academic* ways.

This false analogy becomes especially obvious when we disclose one additional fact about these twins—the twins are only *fraternal* twins, with one being a male and one being a female. Although they may be similar in many ways, the simple fact that one is a male completely removes any possibility of him becoming pregnant, much less of becoming pregnant at the same time as his twin.

In a similar way, Phoenix and Philadelphia are both major cities. Their populations are nearly the same. But these two similarities cannot be used as a basis for determining how much money the cities will need for highway snow removal this winter. This is clearly a case of *comparing apples to oranges.* That is, both apples and oranges are fruits, and both are round. These similarities, though, are completely irrelevant to predicting whether they are of similar color.

The examples presented so far are obvious cases; however, in many cases, it is more difficult to determine whether the cases are similar enough to warrant similar treatment. For example, the impotence drug, Viagra, is often covered by medical insurance. However, birth control pills for women are often not covered. On the surface it appears that these cases are so similar that they both should qualify for insurance. Although there are many OTHER reasons why birth control pills should be covered, suggesting that they be covered because Viagra is covered is a case of false analogy, or at least, questionable analogy. The cases are different in a key way—Viagra treats a medical condition that prevents people from engaging in normal human behavior. Birth control pills, on the other hand, do not treat a medical condition. Again, although there are several other very strong arguments as to why birth control pills should be covered by insurance, this analogy is probably not one of them.

In using reasons to support an argument, it is important to use strong arguments, not weak ones. Weak arguments, even those that seem to support your case, often end up doing more harm than good. In looking at the issue of Viagra and birth control, the analogy seems to support covering both the Viagra and birth control pills. However, upon closer inspection, opponents of your position will point out the flaw in the analogy and use that as evidence that your overall reasoning abilities are biased or questionable. Instead of trying to salvage an argument that is badly injured, it is probably better to focus on presenting other, stronger reasons. For example, if one goal of insurance is to keep medical expenses down, covering the expense of birth control pills is far less expensive than covering the cost of a pregnancy and birth of a baby.

The first analogy about the similarities of the twins is instantly and clearly recognizable as a false analogy. The Viagra-birth control debate takes a bit more analysis to see why there is some weakness in the analogy. Two people can come to the same conclusion—that insurance should cover birth control pills—but there are correct and incorrect ways of arriving at that conclusion.

Frequently we assume that others who share our political views, for example, came to those conclusions based on the same reasoning that we used. This is often an incorrect assumption. There are logical and illogical ways of arriving at the same conclusion. It DOES matter how we arrive at our conclusions. We should not be especially impressed with those who share our opinions if they arrived at those opinions by using faulty reasoning or by relying on incorrect information. On the other hand, we should value and respect those we disagree with, as long as they can show that they are using logical reasoning and correct information to arrive at their conclusions. Voting for a particular candidate based on careful reasoning and correct information is to be admired. But, we should not particularly admire one who votes for the same candidate simply because of a coin toss. Using a coin to make such a decision is inappropriate in much the same way as using a faulty analogy is inappropriate.

As mentioned earlier, some analogies are clearly strong, and some are clearly weak. Determining the strength of some analogies can often be very difficult. For example, consider the 2008 court decision in California that allows same-sex marriage.[11] Obviously there are good reasons both supporting and opposing this verdict. But one analogy has received significant attention—one concerning the similarity of same-sex marriage and interracial marriage. In the past, in many parts of the United States, laws prohibited individuals from marrying one from a different race. In fact, it wasn't until the 1960s that the Supreme Court actually ruled that any such laws were unconstitutional.[12] Does this ruling suggest that laws banning same-sex marriage are also unconstitutional? Some argue that the analogy is a very strong one, whereas others argue the opposite.

Last, consider the fact that insurance sometimes covers dental braces, which turn crooked teeth into straight teeth. Should insurance companies also cover costs associated with straightening one's crooked nose? Most insurance policies say cosmetic procedures are not covered; therefore, they would disallow a nose job. Should braces also be disallowed because they too are primarily for cosmetic reasons? Or, are there enough differences to question this analogy?

These last couple cases are complex enough that intelligent people can debate both sides and legitimately reach different conclusions. However, clear false analogies occur frequently because individuals often have only a superficial understanding of the items or issues they are comparing. The similarities might be visible and obvious; however, the differences are often substantial enough to make the analogies either false or questionable.

Fallacy

of

Special Pleading

> *Dr. Thielen, I missed the test last class because my Mom didn't wake me up to get here on time. It's okay for me to take it today, right?*
>
> —Actual Statement from a College Student in the Author's Class, 2005—

The *Fallacy of Special Pleading* is a flawed way of thinking, in which an exception to a rule is requested, but no reasonable excuse is offered for the exception.

The *fallacy of special pleading* is related to the *fallacy of sweeping generalization,* which will be explained later. The sweeping generalization is a blanket statement that allows for no exceptions to the rule, not even reasonable exceptions due to very extenuating circumstances. In contrast, the fallacy of special pleading asks for an exception to the rule, but no reasonable reason is provided as to why the exception should be granted.

In the case above, the student believed that his excuse was, in fact, legitimate. Consequently, he believed he had committed no fallacy. The professor, however, suggested otherwise. (Note: In accordance with the syllabus, the student was allowed to take the exam at a later date for reduced credit.)

This fallacy is sometimes known as *having it both ways.* That is, we demand that others abide by laws and rules; however, when we break the law, we suggest that our case is somehow *special.* This really is a fallacy of *consistency.* One of the hallmarks of effective reasoning is to be consistent in applying logical principles to similar cases.

For example, I recently heard an individual complain that the state is borrowing money to fund its obligations. However, not more than a few minutes later, the same individual praised the federal government for borrowing money to fund its obligations. Unless the individual can demonstrate that these two cases are clearly dissimilar, praising one and criticizing another is inconsistent.

In a complicated world, inconsistencies occur frequently. For example, most people demand that the candidate or proposal that wins the majority of the votes should therefore win. However, in many states, a simple majority vote often does not determine the outcome of a particular proposal. How does this occur? In order for certain measures to pass, a supermajority of 66.6% of the voters must favor the proposal. So, if such a proposal is supported by 65% of the voters, a clear majority, it still fails. In essence, just 35% of the voters can defeat a proposal. At least on the surface, the principle of rule by a simple majority is violated.

Fallacy of Sweeping Generalization

Arnold Schwarzenegger, the famous actor, bodybuilder, and now governor of California is obese, according to the standards set up by the Federal Government.[13]

The *Fallacy of Sweeping Generalization* is a faulty way of thinking, in which a general rule is applied to all cases, even to those cases that are exceptional, unique, or abnormal.

A sweeping generalization is an oversimplified, broad statement that begins with a general rule and allows no specific exceptions to that rule. As an analogy, a broom is often used to sweep up whatever is in its path, allowing nothing to escape beneath its bristles.

In the case of Arnold Schwarzenegger, the federal government provides a sweeping generalization by asserting that all people who have a body mass index (BMI) over 30 are *obese*. Although this rule is generally true, those people who are especially muscular often have BMIs above 30, yet can be extremely fit and trim.

Encouraging people to exercise vigorously three times a week is another such example. Although it is generally true that people should exercise vigorously three times a week, those who just suffered a heart attack should obviously be exempt from this rule. Similarly, a professor's policy that allows no student to makeup a missed exam, regardless of the specifics, commits this fallacy. Obviously, missing an exam due to a car accident is a case in which a reasonable exception to the rule should be allowed. Sometimes people make sweeping generalizations because of faulty reasoning. However, sometimes individuals will knowingly provide a sweeping generalization, and include the *no exceptions allowed* phrase, as a warning to those who might think they can easily *bend the rules*. If clear rules and policies are not uniformly and rigorously enforced, chaos often reigns. However, in virtually all organizations, whether it is in a government agency, company, or school, an appeal process allows those who have extraordinary cases to argue that an exception to the rule should be allowed.

A few unethical insurance companies, for example, are notorious for denying medical or automobile claims. Sometimes when an individual makes a legitimate claim, the insurance company will deny it by finding an insignificant detail in the claim that, the company says, makes it ineligible. Furthermore, the insurance company will sometimes substantiate the denial by referring to pages and pages of broad statements and policy documents. That is, no exceptions, not even reasonable ones, are allowed.

A particularly annoying sweeping generalization occurs when one attempts to redeem coupons, rebates, special offers, and other incentives. Notice how frequently these enticing offers turn out to be of no value because rules, often in tiny print, disallow the claims.

In fact, a common advertising tactic of *bait and switch* employs this fallacy. For example, automobile dealerships are well-known for offering *incredibly low finance rates of just 2.9%*, or other such advertising gimmicks. Such a rate, for example, baits or entices customers to come to the car lot. Once there, however, customers often find they do not qualify because the rule says only those with phenomenal credit are eligible. The bait was the finance offer. The switch, of course, occurs when the sales person attempts to keep the customer interested by offering some other option, program, or product.

This fallacy of the sweeping generalization is also known as the *fallacy of accident*. A loosely worded law stating that a *murderer* is defined as one who *intentionally causes a death* may have accidental or unintended consequences. In fact, given this sweeping generalization, a doctor who carries out a federal death penalty order by giving a prisoner a lethal injection could be charged with murder. Obviously, it is an accident or oversight that the law was written in such a way that it does not allow such an exception.

Gula, in his book *Nonsense*,[14] sums up this fallacy eloquently by stating simply that this fallacy occurs when the letter of the law is applied, but the spirit of the law is ignored. Most rules and generalizations are intended for covering normal cases, but applying a blanket statement in all cases, even highly exceptional ones, is usually unreasonable.

Chapter 4

Fallacies of Causation

Cum Hoc

Post Hoc

Single Cause

Reversing the Cause and Effect

Ignoring a Common Cause

Texas Sharpshooter

Regression to the Mean

Gambler's Fallacy

Superstition

Chapter 4

Fallacies of Causation

Georgia Gets Rain Day after the Governor Leads Prayers to God

This headline: *Georgia Gets Rain Day after the Governor Leads Prayers to God,* appeared in dozens of newspapers across the country. Georgia, like many parts of the Southeast, was drought-stricken in the summer and fall of 2007. The Governor, Sonny Perdue, led a prayer service on the steps of the state capital to *beg the heavens to end the drought.* The next day the state was drenched in rain. Later Perdue stated, *Frankly, it's a great affirmation for what we asked for.*[1]

To many people, this story is proof of divine intervention. That is, rain was prayed for and then a divine being *caused* it to rain. Such people would say that this is a clear case of cause-effect. In other words, the cause was prayer, and the effect was the rain. Is it really that simple?

Although it is certainly very comforting to believe in such a simple notion, critical thinkers, even those who are highly religious, warn us not to let blind faith overpower logical reasoning. What else, other than a divine being, could cause the rain? What explanations can science and critical thinking provide?

This example serves as a perfect introduction to the fallacies of causation. Simply put, a fallacy of causation occurs when someone identifies an incorrect or inadequate cause of some event. Believing that washing your car causes it to rain, or finding a penny causes good luck are some of the most obvious and superficial examples of this fallacy.

Why am I sick? What caused me to become sick? What are the side effects of the medicines I am taking? These questions, and more importantly, the answers to them, all deal with the issue of causation. The fallacies of causation often produce wrong answers to these, and many other questions.

As mentioned in the introduction of the book, people have a thirst to know why certain events happen. Even when the *why* is unknowable, the mind will literally manufacture reasons to explain why something happened. Despite a complete lack of proof, the mind will often latch onto a reason because, as the quote says, *it sounds good*. The mixture of pseudoscientific and bizarre reasons that people accept in their attempt to explain why certain things happen is almost beyond belief.

There is one limitation of this book that readers should know—it makes no attempt to deal with religious beliefs. Most such beliefs are not typically based on scientific evidence, but based on faith. In fact, faith is sometimes defined as *belief that does not rest on logical proof or material evidence.*[2]

In contrast, the point of this book is that logical proof and material evidence are the foundation on which high-quality *knowledge* is created. But, *knowledge* is not the same as *belief.* Those who say they *know* something should be bound by the principles of critical thinking. This distinction is often made clear when people talk about God or some higher power. Many will say that they *believe* that God exists, but that they do not *know* that God exists. *Knowledge,* at least as far as a critical thinker is concerned, develops from evidence and rationality. As long as people acknowledge that the vast majority of religious beliefs are beyond the realm of rationality, there is not an issue to debate. *Rational knowledge* develops when evidence and reason are used in a way that meets standards of proof or quality. In contrast, *beliefs* do not have to meet any such standards. It is for this reason that the vast majority of credible scientists *know* that earthquakes are caused by shifting plates on the crust of the earth. This is *knowledge,* because it has met the standards of science. In contrast, those rare individuals who suggest that an angry God causes earthquakes are merely expressing a personal *belief*—an opinion not based on logical reasoning or evidence.

Obviously knowledge can lead to beliefs, but an effort should be made to keep this distinction between knowledge and belief in mind. For example, in considering what caused life to appear on earth, science attempts to provide answers based on knowledge, whereas some religions answer this question with a belief.

Nothing in this book on critical thinking should be construed as being opposed to religious philosophies or a belief in God. Quite the contrary is true; it respects these beliefs. A belief in God, for example, serves many useful functions. For example, medical science frequently cannot offer hope to a patient who is dying. However, a belief in something beyond rationality, or a belief beyond medicine, can provide hope. And it is this hope—or belief, or faith—that is responsible for maintaining a positive attitude, an attitude that may be an effective antidote for some illnesses.

Of course, one should be careful in assuming that God was necessarily responsible for some event or recovery that is beyond the expectations of current medical or scientific knowledge. For example, 50 years ago medical science assumed that the brain could not recover from being deprived of oxygen for more than a few minutes. In those rare cases where an individual apparently drowned and *seemingly came back to life*, the claim was that a *miracle* occurred. In fact, perhaps a *miracle* did occur. On the other hand, medical science can now explain why some of these cases occur. In those cases where someone is underwater for more than several minutes, as is the case when someone might fall into an icy cold lake or river, the body temperature may fall rapidly and therefore allow the brain to survive much longer than normal without oxygen.

Similarly, one should be careful in assuming with certainty that a supernatural force, God, or a guardian angel, was responsible for our avoiding a tragedy that struck almost everyone else. The fact that only one person avoided being killed in a plane crash does not necessarily mean that a *guardian angel* protected that individual. With this reasoning, one could logically ask why none of the others were saved by their *guardian angels*.

In fact, instantly assuming that some supernatural force is responsible for events that are unexplainable is often considered a fallacy in its own right. Many of the modern conveniences that we now enjoy, such as electricity, microwave ovens, remote controls, nuclear energy, airplanes, radio, computers, telephones, cameras, refrigerators, and television would have been considered to be miracles, beyond rationality, or supernatural, just a couple centuries ago. Again, not being able to provide a logical explanation for something is not hard evidence of a supernatural influence. In the same way, the fact that we are unable to explain many magical tricks or illusions does not suggest that supernatural forces are at work.

Despite these issues, religion and a belief in God, do not necessarily conflict with critical thinking. Some of the most respected scientists in the world are highly religious. Science is a tool that helps us to explain, for example, how life began and evolved. It cannot, however, determine why life exists or the meaning of life. These are issues beyond the realm of science and critical thinking. They are, in fact, issues of belief and faith.

Francis Collins, Ph.D., a physician and the director of the National Human Genome Project, helped to map more than 20,000 human genes. He is also a man of strong faith, religion, and belief. Unlike some scientists, he asserts that one can believe in both God and science. A Gallup Poll in 2008 showed that 44 percent of Americans believe God created human beings in their present form within the last 10,000 years; however, Collins says: *You can't arrive at that conclusion without throwing out all the evidence of the sciences.*[3] Like Collins, critical thinking DOES refute some extremely fundamentalist religious notions, such as the belief that humans were created in their current form, or that the world is just 6,000 years old. It does not, however, refute one of the most fundamental tenets of most religions, a belief in God. Nor does it refute those who do not have this belief.

For this chapter on causation fallacies, we will limit our study to how rational, scientific, and psychological elements cause various events or consequences to occur. Of all the fallacies in this book, those that deal with faulty causation are extremely common and especially damaging.

Fallacy

of

Cum Hoc

I heard about a study that says people with long hair are far more likely to get pregnant that those with short hair. I need to let my hair grow out because I want to have a baby.

The *fallacy of cum hoc* is a faulty way of thinking, in which one automatically assumes that one event caused another event simply because the events occurred together or at the same time. *Cum hoc* is actually short for *Cum hoc, ergo propter hoc*. This term comes from Latin and means *with this, therefore because of this.*[4]

One of the most basic principles in science states that *correlation does not mean causation.* College students have probably heard this dozens of times; however, few really understand exactly what it means. Our intuition tells us that when two events occur together, one event is usually the cause, and the other event is usually the effect. But, our intuitive belief in this cause-effect relationship is often wrong. With enough data, and enough analysis of that data, all kinds of relationships can be found, but often these are simply coincidental relationships.

In our example above, the relationship between long hair and higher rates of pregnancy is true. However, one doesn't cause the other; it is simply a correlation. This relationship can be easily explained by the fact that women, in general, tend to have longer hair then men. Women also can become pregnant, unlike men.

Such relationships are common. For example, federal expenditures in the United States during 2000 through 2006 went up dramatically. At the same time, the number of facelifts also went up dramatically. Obviously, this relationship is not causal. Math SAT scores went up significantly from 1995 to 2000.[5, 6] The U.S. stock market also went up significantly at the same time. Again, this relationship is coincidental. The number of computers in the U.S. increased dramatically from 1970 on. At the same time, the number of divorces also increased dramatically. Again, the first event did not cause the second event.

Most people instantly understand, by looking at the previous examples, that assuming a cause-effect relationship would be ludicrous—simply because the suggested correlations are so outrageous and extreme. However, the same skepticism must be applied to examples that seem more plausible.

A presidential candidate, for example, states that, while he was governor, crime in his state went down. A health-food manufacturer says that, as more people drink bottled water, life expectancy has increased. In both cases, two events occurred at the same time. But these links, without proof of causation, are simply correlations.

Hundreds of such relationships are brought to the public's attention each year via the mass media, and often the mere suggestion that one event caused the other event is enough evidence for some to proclaim that the apparent cause-effect relationship is a *fact*. These errors occur frequently when it comes to the conclusions that people assume from medical or nutritional studies. Researchers are typically quite careful in pointing out that any relationship they initially find between a particular food, for example, and a particular disease is simply a *link* or a *correlation*. However, millions of people incorrectly believe that saying there is a link between two items is the same as saying there is a causal relationship. And, as mentioned previously, simplistic thinkers often believe passionately in the *truth* of their own conclusions and then spread this misinformation to others.

There are several fallacies that deal with causation, but the distinguishing element of this fallacy centers on the fact that the two events occurred *together, at the same time*. In contrast, the next fallacy we will examine fools people into believing that one event caused another event simply because one occurred *after* another event.

Fallacy

of

Post Hoc

> *I took the pills the doctor gave me, and the next day my stomach was upset. It is pretty clear to me that the pills caused my stomachache.*

The *fallacy of post hoc* is a faulty way of thinking, in which one automatically assumes that because one event occurred *after* another event in *time*, the first event *caused* the second event, when in fact, it may have been merely a coincidence.

Post Hoc is actually short for *Post hoc, ergo propter hoc.* This term comes from Latin and means *after this, therefore because of this.*[7] Just as in sports, the *post* in the *postgame analysis* obviously occurs *after* the game.

In our example above, the first event of taking the pills is believed to have caused the second event, the upset stomach. Although it is very tempting to believe this, it may not necessarily be true. In fact, the initial assumption should be that it did *not* cause the upset stomach. Just as we assume that someone is innocent until proven guilty, we should also assume, without strong evidence, that the first event did not cause the second event. A simple time relationship is not strong evidence of cause-effect.

Because the time between taking the pills and having an upset stomach is so short, there is a strong intuitive belief that one event directly caused the other. Although it is possible that one side effect of the pills is an upset stomach, it is conceivable that the very illness or condition for which the pills were prescribed is causing the upset stomach. If the pills were prescribed for a stomach ulcer, perhaps it is really the worsening ulcer that is causing the stomachache, not the pills. Eating a particular food, riding in a car, or becoming emotionally upset may have occurred soon after taking the pills—they may be the real reasons for the upset stomach.

Isolating the precise cause of a particular health symptom or side effect is often extremely difficult—especially when one may be taking various medications at the same time. In some cases, whether a particular effect was caused by medicine, lack of sleep, change in the weather, physical stress, or any number of other daily variables is extremely difficult to determine. Often a patient will experience some negative effect in the days after taking a new medicine and instantly believe it was the medicine that caused the negative effect. Consequently the patient will demand that the doctor write a different prescription, when there was really another cause for the negative effect. Like a technician who troubleshoots a computer problem by locating the particular bug through a systematic process of elimination, most people would benefit from trying to apply similar thinking to many of their own problems.

There are several fallacies that deal with causation, but, as mentioned above, the distinguishing element of this fallacy centers on the *timing* of the two events. In other words, the chronological ordering of events is assumed to be justification for believing in a cause-effect relationship. Using time as one *possible* indicator of a cause-effect relationship is not incorrect, but those who use time, and only time, as definitive proof of a cause-effect relationship are guilty of this fallacy.

There are many examples in which some event follows another event simply because of coincidence. Washing your car suddenly *caused* it to rain. Stepping in the shower *caused* the phone to ring. These are obvious examples of post hoc; however, there are many examples that are far less obvious.

For example, there is a clear relationship between my headaches and seeing my mother-in-law. Whenever I went to her house, I would usually get a headache after an hour or so. Did she *cause* the headaches? Some people would arrive at such a conclusion, because it occurred so often and because there was such a short time between the first event and the second event. However, after some investigation, I learned that my mother-in-law was not causing my headaches. Instead, excessive caffeine was the culprit. At her house, whenever I arrived, almost always the first thing I would do is to drink a few cups of coffee, coffee that was always readily available. My excessive coffee drinking caused excessive caffeine consumption, which caused my headaches.

In theory, there are many other possible reasons. Perhaps I was allergic to her cat, or perhaps to her perfume. In fact, there could be dozens of reasons to explain this relationship. Again, pointing to just a temporal or time relationship between two events is not adequate proof that one caused the other.

I noticed, for example, that I lost some weight in the few weeks after I began wearing a new pair of glasses. Did the first act of wearing the glasses cause the second event of losing weight? The glasses were actually the wrong prescription, and they caused me to feel slightly nauseous. Because I was nauseous, I ate less and therefore lost some weight.

People have pointed out that after fluoride was put in city water supplies across the country, cancer rates increased. It is a fallacy to assume that fluoride causes cancer based solely on this temporal relationship. Similarly, it was found that almost all children who attend school develop some tooth decay. Does going to school cause tooth decay? Prior to *every* divorce, a marriage occurred. So, marriage is the *cause* of divorce?

Identifying the specific social, economic, or political causes for historical events is very difficult, and it often invites simplistic and dangerous post hoc reasoning. Claiming that more violent song lyrics in the 1970s were responsible for more violence on the streets in the 1980s also employs simplistic post hoc reasoning. In some parts of the country, murder rates went down after the death penalty was reinstated. Simply assuming that reinstating the death penalty is the *cause* for fewer murders is another such example of the post hoc error.

Even many so-called obvious cause-effect relationships often become questionable under scrutiny. For example, everyone *knows* that eating foods that contain substantial fat leads to high levels of cholesterol. Consequently many people have correctly decided that a low-fat diet is healthy. And ongoing research continues to support the merits of a low-fat diet. But, are there exceptions to this general recommendation? Can a diet that includes a higher level of protein and fat also be healthy? The Atkins Diet, or the low-carb diet, for example, has been criticized for years because of an assumption that a diet allowing fat would instantly cause cholesterol problems.

Surprisingly, the vast majority of studies have found the opposite.[8] The most recent study, published in the *New England Journal of Medicine* (July 17, 2008) is hailed by experts as highly credible because the study lasted more than two years, used a large number of people, and was conducted under highly controlled conditions. It was one of the longest and largest studies comparing diets. Although this study found that low-fat diets were helpful in obtaining ideal cholesterol levels and losing weight, it also found that the Atkins diet did just as well, or better than, all of the other diets studied on these measures.[9]

So, the initial and *obvious* cause-effect relationship that more fat causes more cholesterol may not necessarily be as solid as once thought. Medical and nutritional research, like most research, goes through a process of isolating causes and effects. Often a conclusion from initial research will go through a more rigorous process of evolution in which it is corrected, adjusted, or refined. For example, 40 years ago, high cholesterol in general was considered a major problem. However, we now know that having a high level of HDL cholesterol is desirable, whereas having a high level of LDL remains undesirable.

There are obvious relationships that are clearly not cause-effect ones. But, more importantly, there are millions of other links or connections that might seem to have some plausible or intuitive cause-effect association. There are many such relationships that especially demand our skepticism. Do fewer murders, for example, suggest that violence is going down? Or, does it mean, for example, that the ambulances are arriving faster to save those who have been shot? Or does it mean that emergency room doctors are more skilled and have more effective methods of saving lives than in the past? Similarly, teachers are finding fewer drugs on some campuses. Does this mean that drug use is going down or that students are better at hiding the drugs?

It is not surprising that the masses of people search for simplistic time-related cause-effect relationships where none truly exist. Even scientists and researchers admit that it is extremely difficult to isolate the exact causes of many events.

Fallacy of Single Cause

> *Taxes were cut a few years ago, and now the economy is in great shape. It is quite simple—the tax cut is the reason our economy is in such great shape.*

The *fallacy of single cause* is a faulty way of thinking, in which one identifies a single cause for an event when, in fact, two or more causes are responsible.

In the example above, most economists agree that a tax cut can stimulate the economy, at least on a short-term basis. However, the vast majority of economic, political, or social effects are caused by several factors.

Besides the tax cut, what other factors may be just as important? During this time, interest rates by the Federal Reserve were low, and low interest rates usually enhance economic activity. Spending by the federal government, although much of it was borrowed money, also increased. This too typically increases economic performance temporarily.

In an attempt to look for simple solutions or explanations to complicated problems, one's natural tendency is to accept a single explanation, not a rash of complicated theories or complex solutions. Politicians know that voters are often comforted when a single cause is identified as the reason for a problem.

Law enforcement officials may contend that the single reason that violent crime declined in the 1990s is that there were more officers put on the street. On the other hand, judges might suggest that stiffer sentencing laws were the reason for the decline. A president may claim that the single reason crime went down during this time was due to federal economic policies that lowered unemployment, which in turn, lowered violent crime. It is very likely that all of these factors had some influence on the rate of violent crime.

Another very important but little publicized variable that probably affected crime rates was demographic changes. Those who are 15 to 45 years old commit the vast majority of violent crimes. There were more than 75 million births in this country from 1946 to 1964, the baby-boom generation. During the 1960s, 1970s, and 1980s, large numbers of people were precisely in this age range. But, by the early 1990s, millions of these boomers started getting old (exceeding 45 years old). In other words, by 1990, the number of people in the *prime violence age group* began declining. That is, the pool of likely criminals was much smaller during this time. Is this part of the reason why crime declined?

So what caused the declining violent crime rates in the 1990s? More police on the street, harsher sentences, low unemployment, and changing demographics, are all variables that need to be considered.

Those who commit the *fallacy of single cause* often suggest that the reason no terrorist attack has occurred in the United States since 2001 is because the terrorists are being fought in Iraq. While this is conceivable, there are many other plausible reasons as well. Perhaps improved airline security measures or stricter border control policies are the cause. Perhaps terrorists are now concentrating their attacks against countries that are allied with the United States. Perhaps some of the planned attacks were thwarted by improved intelligence or by pure luck. Perhaps the terrorists have no great desire to launch another immediate attack because the first attack created the media attention and economic damage they had hoped for.

Based on information gathered about the 2001 attack, it is clear that some terrorist organizations spend many years planning their attacks. Could it simply be that the next major attack is still in its planning stages? Avoiding the *fallacy of single cause* demands that all of these factors be considered.

Fallacy

of

Reversing the Cause and Effect

I'm going to quit drinking diet soda. I thought drinking it would help me stay thin. But, they found that more people who drink diet soda are fat than those who drink regular soda.

The *fallacy of reversing the cause and effect* is a faulty way of thinking, in which one mistakes the cause as the effect and the effect as the cause. This is also known as the *fallacy of wrong direction.*

Although it is tempting to believe that the cause and effect are usually very obvious, this is often not true. One notices a relationship between drinking and depression, and assumes that the drinking causes depression. But, one could conclude just the opposite—that depression causes people to self-medicate by drinking alcohol. Most medical research suggests that alcohol often acts as a depressant, and it probably does induce depression in some people. However, some research indicates that those who have a propensity towards depression seek ways of relieving their depression, such as by drinking.[10] This drinking may temporarily lessen depression; however, long-term it usually makes the depression worse.

So, the relationship between depression and alcohol may be cause-effect or effect-cause. In other cases, the mix-up is far more obvious. There is a clear relationship between lightening and thunder. As mentioned previously, one cannot conclusively say that lightening causes thunder based *solely* on the chronological order in which they happen. Besides the timing of the events, scientists use other methods to verify the causal relationship. Reversing this relationship, in this case, is not a mistake people normally make. That is, no one claims that thunder causes lightening.

In our previous example, one should not conclude that drinking diet soda is the cause, and obesity is the effect. In fact, being obese probably influences one to drink diet soda as a way of preventing even greater obesity.[11]

The relationship between vigorous exercise and weight can also be analyzed. While those who exercise and eat healthy are more likely to keep their weight under control, one should not be overly impressed with the many gorgeous and slender models featured in TV commercials promoting a particular diet, exercise program, or fitness center. Obviously the advertiser is trying to persuade us to believe that working out at a gym or following a particular diet will make us more physically attractive. This is often logical and reasonable. However, it is logical and reasonable to assume that those who appear in the TV commercials are sometimes already attractive and slender. Indeed, it may be that their initial attractiveness is the cause for their selection to appear in TV commercials in the first place.

A ceiling fan can run in reverse, and the change in direction may still provide some benefit. In contrast, one cannot reverse the cause-effect direction in an argument and assume it is still valid.

Fallacy

of

Ignoring a Common Cause

> *A study found that women who have facelifts actually live longer than those who don't.*[12]
> *So, I think I am going to get a facelift myself.*

The *fallacy of ignoring a common cause* is a faulty way of thinking, in which one automatically assumes that one event caused another event when, in reality, both events were caused by a third event.

In other words, one notices a correlation between two events and thinks one event is the *cause* and the second event is the *effect*. However, *both* of the events are actually *effects*, and a completely *different event* is the *cause of both prior events*. In our example earlier, one is assuming that a facelift will cause a longer life. Upon more careful reflection, however, one should realize that a longer life and a facelift are both effects caused by a different event, variable, or cause. The cause that is common to both events is better medical care. That is, those who have facelifts tend to be wealthier, and those who are wealthier tend to have better medical care that results in a longer life.

Other examples are far more obvious. For example, I have noticed that there is a clear relationship between my allergies and getting a tan. When I get a tan, almost always my allergies start to flare up. Assuming my tan actually causes the allergies is a fallacy. Both my tan and the allergies are caused at nearly the same time because the sun causes them both. More specifically, the rays of the sun cause my tan, but the heat of the sun causes warmer temperatures, which cause various plants and flowers to bloom. The blooms, then, activate my allergies.

The *Internet Encyclopedia of Philosophy* (2009) illustrates this fallacy by pointing out that automobile accidents increase when windshield wipers are used.[13] Assuming that the wipers cause the accidents is obviously the flaw. Both using the wipers and the increased accident are effects, both of which are caused by some common element. That element, of course, is the rain.

Similarly, the sun may be the common cause of other related events. For example, warm weather and drowning are usually correlated—as the temperature increases, the number of people who drown increases. However, warm temperatures don't really cause more people to drown. The warming temperature of the summer sun, for example, causes more people to go swimming in their pools or at the beach. Naturally, more people in the water increase the probability of more people drowning.

Many health issues are difficult to diagnose because two symptoms are noticed, with one labeled the cause and the other labeled the effect. An ear infection, for example, may appear to be the cause of a fever; however, it could be that both the ear infection and the fever actually resulted from a virus.

I, like many other professors, am guilty of suggesting a relationship between college and financial success that is not as absolute as many people think. For example, those who earn a college degree have significantly higher salaries, on average, than those who don't have college degrees. Is this difference, commonly reported at nearly $20,000 a year[14] (over a million dollars during an average lifetime), caused by college? Probably much of it is—college develops skills that employers are willing to pay for. However, could it be that *some* of the salary difference is due to the fact those who choose to attend college in the first place already have some of these skills? Could it be that *both* getting a degree and financial success are *effects,* and that a partial *cause* of both of them is initial motivation, intelligence, or ability?

Similarly, many colleges that charge outrageously high tuition suggest their fees are *worth it* simply because their graduates make more money than the typical college graduate. However, often those attending highly competitive and expensive colleges are already more intelligent or skilled than the students attending less expensive state schools that have lower entrance requirements. Substantial evidence indicates that such able and intelligent students, regardless of whether they attend expensive or inexpensive colleges, will earn higher salaries than the typical graduate. That is, the more expensive education is not necessarily the primary cause for higher salaries.[15]

In fact, *studies indicate that, for equally able students, education at a highly selective college does not lead to a better job or higher income than does education at a less-selective college.*[16] Despite opinions to the contrary, generally the major variable to future success is the student, and not the college he or she attends.

On a lighter note, there is an often-told story that begins with a man who drinks six *whiskey and cokes* and then he complains about having a headache the next morning. A few days later he drinks six *rum and cokes* and then again complains about having a headache the next morning. Finally, several days later he drinks six *vodka and cokes* and once again complains about having a headache. In searching for a common cause for his headaches, he concludes that the *coke* is causing the headaches. Obviously, the *coke* is not causing the headaches, but rather the alcohol in the drinks is.[17] This is a rather obvious example of the *common cause fallacy;* however, in many other situations, it is exceedingly difficult to determine whether there is another unknown variable that is causing the symptoms. This problem is especially common in medicine.

Like the other causal fallacies, this one entices us to see a relationship between two events, when, in fact, a third variable is operating. Taking the time to suspect that there might be a third variable operating is the first step in critical thinking. Finding that variable is often extremely difficult.

Fallacy

of

Texas Sharpshooter

In the southwest valley last year there were 25 premature babies born that weighed less than three pounds. But, in the rest of the entire valley there were only ten such cases. That nuclear power plant is right in the middle of the southwest valley, and I think the plant's radiation is affecting a lot of pregnant women. That has to be causing many of those premature births.

The *Texas sharpshooter fallacy* is a faulty way of thinking, in which one concludes that a clustering of cases or a streak of events is caused by some specific external factor when, in fact, the cluster is due to random chance. Another aspect of this fallacy acknowledges that these clusters are often simply illusions created by the way people organize information.

This fallacy, which is often called the *clustering illusion,* is frequently found in epidemiological studies when researchers find diseases or physical ailments clustered together in the same locations.[18] When such a cluster of ailments is noticed, many people commit this fallacy when they assume that the cluster was caused by some environmental factor such as polluted groundwater, radiation from a nearby nuclear plant, toxic farm chemicals, or other such dangers. Intuition often tells us that a nearby dangerous element caused these health-related clusters. However, that intuition is often wrong.

Some events will cluster together simply by chance when there are enough events to examine. Winning streaks or losing streaks occur frequently when enough data is reviewed. The so-called *hot-hand* or *cold-hand* in basketball seems to defy the odds; however, such streaks are normal. In fact, as will be shown later, such streaks can be predicted with significant accuracy.

In the example above, suppose a city is divided into four quadrants. The fact that one quadrant had 25 cases of premature births, and the other three put together only had 15 cases might cause one to ask whether something in the southwest part of the city is causing the defects. Upon statistical analysis, though, such a concentration of cases is not unusual. Of course, all cities won't individually have that type of clustering, but when 50 cities, for example, are examined, these clusters will occur on a regular basis.

Rather than using a city for analysis, suppose a large checkerboard of 16 squares is drawn on a large driveway, and assume someone is then given two hands of loose quarters. Assume further that this person is asked to throw these quarters (assume 100 quarters) as high into the air as possible, letting the quarters fall back down on the driveway at random. Would exactly 6.25 quarters fall on each of the 16 squares? On average 6.25 quarters would fall on each square, but some would have more and some less. With that many quarters, it is likely that one square might end up with more than 10 quarters while another square might have only one or two. The fact that one square might have five times as many quarters as another square does not mean that the high concentration of quarters on that one square was caused by pollution, toxic chemicals, or any other environmental factor. The variation was simply caused by chance.

Before investigating this fallacy further, looking at the origin of this thinking error might help to illustrate how it can occur. The *Texas sharpshooter fallacy* derived its name from a story about a Texas cowboy who took his gun and randomly fired several dozen shots into the end of a barn. Then he examined the end of the barn and located the area that had the heaviest concentrations of bullets. He then painted a bulls-eye target around this cluster, which made it appear to others that that he was a skilled sharpshooter. So, given enough shots, or cases, one can make it appear that the clusters or streaks were caused by something other than random shots.[19]

In speaking about randomness and streaks, there are a number of results that are logical, yet seem counterintuitive. I have had scores of classes over the past ten years perform the following experiment. Every student in each class of 25 is asked to flip a coin 20 times and record on paper the result of each coin flip, heads or tails. Once all the students had completed this exercise, the resulting data showed that the probability of having a streak of at least four heads in a row within those 20 throws was 50 percent. The probability of five heads was 25 percent. And, despite what most people believe, the odds of having a streak of at least six heads in a row was 10%. In other words, about 12 from each class of 25 students had streak of at least four; six had a streak of at least five. And two had a streak of at least six heads. These classroom results are exactly what probability theory predicts. Again, random streaks or clusters occur quite frequently, whether they are in the form of bullet holes found on the end of a barn or shots taken by a basketball player.

Thomas Gilovich, Ph.D., in his book, *How We Know What Isn't So* (1991) provides evidence that the so-called *hot-hand* or *cold-hand* in basketball can be easily explained with simple statistics.[20] In fact, one can take the results of my classroom experiments and apply these statistical probabilities to basketball.

Suppose your favorite player typically takes 20 shots per game and normally makes 50% of those shots. How often during the season will your player hit four in a row, or five in a row, or six in a row? He will hit four in a row on a very regular basis, actually in 50% of the games. On average, five in a row will occur 25% of the time, and six will occur about 10% of the time. In fact, with the sheer number of players in the league, and the sheer number of attempts, frequent reports of *hot hands* and *cold hands* should not surprise many fans. Both the *hot-streaks* and *slumps* that individual athletes and teams experience are completely normal.

The importance of this fallacy becomes clear when the masses of small-time individual investors start to listen to a so-called *investment guru*. This is a person who, for example, accurately predicted whether the stock market would be up or down for each of the last ten years. A magazine then implies that this guru is someone we should pay attention to and invest our money as he suggests, because he was right the past ten years. Hundreds of millions of dollars are lost by millions of investors because they fall for this fallacy. There are thousands and thousands of financial advisors who make predictions about the market all the time. There will be several dozen such advisors, from those thousands, who will be right due to pure chance. In other words, if thousands of advisors simply flipped a coin each year to make a prediction, a handful would be right for several years in a row.

In fact, if about 100,000 financial advisors make a prediction as to whether the market will be up or down for each of the next ten years, it is very likely that nearly 100 of them (97.65 to be exact) will provide predictions that will turn out to be absolutely correct—all due to nothing more than chance. Statistically, one person out of 1,024 will pick all ten years correctly. From 100,000 advisors, these 100 may be hailed as those with special financial or predictive abilities. In fact, these superstars are simply the result of statistical probabilities. Again, when enough people or data are examined, streaks or clusters are normal. (Note: This is not to suggest that all, or even most, financial superstars are simply the result of statistical chance. Some have shown their competence in other verifiable ways).

This fallacy points out that streaks or clusters are normal and often simply due to chance. But, another aspect of the fallacy shows another reason why clusters seem to exist. The way in which a researcher, for example, looks at the data is responsible for creating a cluster that does not really exist. For example, look at the locations below of reported cases of muggings on the map of a hypothetical city.

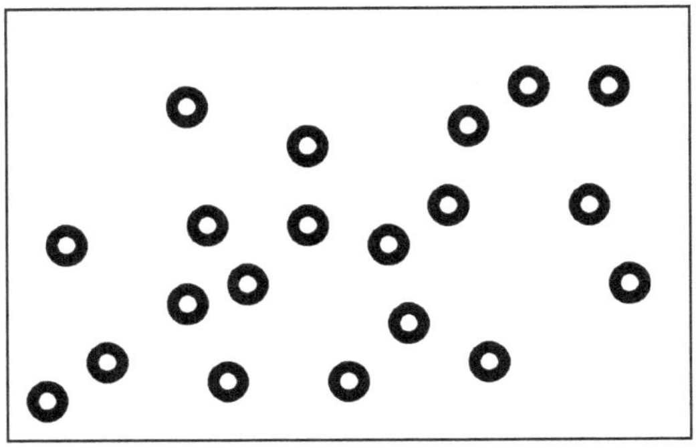

Most people would say the muggings are randomly occurring around the city, and no particular area is more or less dangerous than other areas. That is, no cluster exists.

However, suppose one then marks out the area around the river that runs through the city. Note that the following map is no different from the first in terms of the actual locations of reported muggings.

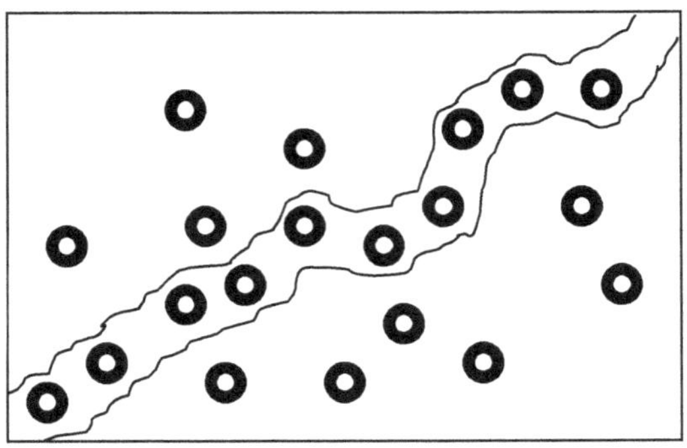

Now it seems that the area around the river is especially dangerous. However, this cluster is created simply by how someone sectioned the city.

In review, clusters and streaks are often not caused by any particular factor other than random chance or probability. The illusion of a cluster can also be created based on how people organize information. This fallacy is a causation fallacy because, like all other fallacies in this chapter, it shows how individuals can easily identify the wrong cause for some event.

Fallacy

of

Regression to the Mean

> *I know the New York Giants are going to lose the Super Bowl (2008) next week in Glendale, Arizona. Their quarterback, Eli Manning, was just put on the cover of* Sports Illustrated, *and now he is jinxed. The* Sports Illustrated Jinx *is a well-known fact; athletes or teams appearing on the cover are doomed to failure. Two weeks ago, Green Bay quarterback, Brett Favre, was on the cover, and, sure enough, they lost this week. In fact, so many famous athletes have refused to be on the cover that they put a black cat on the cover (2002), along with the title, The Cover That No One Would Pose For.*[21]

The *fallacy of regression to the mean* is a faulty way of thinking, in which an event that occurred simply due to the *law of averages* is, instead, believed to have been caused by some other variable or event.

The *law of averages* simply means that most events that seem to be out of line eventually return to normal. Given enough time, most things return to normal. As was explained in a previous fallacy, if enough information is examined, it is common for streaks or clusters to occur that do not seem to be normal. But, in reality, these so-called abnormalities are actually quite normal. The fallacy of regression occurs when these apparent abnormalities begin to return to normal, and something other than the law of averages is believed to have caused it.

In looking at what regression to the mean actually implies, *regression* simply means falling back or returning to its original or normal state. *Mean* refers to a mathematical average. So regression to the mean occurs when events that are now occurring with abnormal frequency, simply return to a normal frequency.

Some events are very predictable—we know a full moon will appear every 29.5 days. So, if a full moon occurred on the first of this month, we know that it will occur again at the end of the month. An earthquake that happens to occur on the 29th is not the cause for the full moon on the 30th.

Obviously, most events don't occur with the regularity or predictability of a full moon. However, by looking at the historical frequency of events, we can calculate an average. For example, one knows that it rains, on average, once a week in some particular area of the country. Then, suppose there is a period in which there has been no rain for six weeks. Given the damage caused by this six-week drought, suppose a local group of residents prays for rain, or perhaps does a rain dance, in the hopes that this will cause it to rain. In the case cited at the beginning of the chapter, it rained the day after the prayers were offered. What caused the rain? Regression to the mean, the fact that rain was due to occur, provides a logical explanation.

In our earlier case with the football team, we know that the favorite team wins, on average, 75% of the time; the underdog wins 25% of the time. Given these averages, events that depart substantially from these norms will probably be countered by events that bring it back to its mathematical norm or average. If the favorite team wins three out of four times on average, and the favorite team has actually won ten times straight, it is likely that the streak will end, and that, over the long-haul, future events will bring the winning percentage back in alignment with a 75% average.

On the other hand, if the underdog normally wins 25% of the time, and they have not won in the last 12 games, it is likely that this losing streak will not continue indefinitely. That is, they are due for a win. Keep in mind though, that the law of averages does not say a particular event is about to happen immediately, but only that events will even out over the long run. The *gambler's fallacy,* which will be explained later, explains that a streak of nine heads in a row does not affect the probability that the next throw will be tails. It simply says, over time, a 50/50 ratio will occur.

In our *Sports Illustrated* example cited earlier, although it is true that hundreds of athletes who have appeared on the cover in the last 50 years have lost games soon after, there is no evidence that a jinx or curse had anything to do with these events. In fact, the law of averages predicted that the number of such performances was entirely normal and predictable.[22, 23]

Contrary to the jinx, the Giants actually won the 2008 Super Bowl after appearing on the cover of *Sports Illustrated,* despite being 12-point underdogs. Does this suggest that the curse doesn't exist? Ironically, some may now say that being on the cover is actually a *lucky charm*. Such a win supports *neither* the jinx theory *nor* the good-luck theory. Simply put, the law of averages, or regression to the mean, was operating. If the sports writers have a long record of being right 75% of the time, it is likely over the next several years that they will continue to be right about 75% of the time and wrong about 25% of the time. In this case, they were wrong. Blaming some curse, a full moon, or rain for the outcome is common; however, statisticians know that that a simple understanding of the law of averages is one of the best predictors of future events.

Suppose the athlete is a professional basketball player who has been in the league and has an eight-year record of scoring 20 points a game, a worthy feat, but certainly not an astonishing one. Then, suppose this player has a season in which his scoring has jumped dramatically to 40 points per game. This incredible surge from 20 to 40 points per game is exactly what attracts the attention of sports magazines. Featuring such a player on its cover sells magazines.

Keep in mind, however, that the average performance of this player is at 20 points per season. The probability is that this player's scoring performance will fall back to his average. It is not likely that such a remarkable season will be repeated for an extended period of time. So, regardless of whether one appeared on the cover of a magazine or not, a fantastic year of averaging 40 points per game is likely to be followed by many more years that are average or even below average. Even when the player scores 30 points per game the next season, many fans will see this as a *failure,* or the *downfall* of the athlete. Many fans fail to understand that the player is simply starting to return to his normal performance. Sometimes they may attribute the decline to reasons that may be plausible, but at other times they may cite a jinx, curse, or some other ludicrous notion.

This fallacy of regression to the mean is a major reason why many investors lose money in the stock market. In fact, failing to understand this fallacy leads many investors to act in ways that virtually guarantee a loss. Rather than *buying low* and *selling high,* they actually *buy high* and *sell low.*

Suppose an investor is shown that over the past ten years, on average, a particular stock increased in value by 10% a year. He then notices that the stock went up 100% in the last two years. Believing that this incredible spike is becoming the new norm, he invests $100,000. Then the law of averages reappears, and within two years his $100,000 of stock plummets to just $25,000. In other words, a stock that was averaging 10% a year skyrockets for a couple of years and nose dives the next few years, leaving the investor with just $25,000. The investor loses $75,000 in two years.

Millions of investors, especially those who invested in the dotcom technology companies from 1995 to 2000, did precisely what this investor did. The chairman of the Federal Reserve, Alan Greenspan, repeatedly warned investors as early as 1996 that they were ignoring the law of averages, and were being *irrationally exuberant.*[24] The law of averages returned by 2000, wiping out trillions in investments and leaving massive losses for investors.

Just as a side note, keep in mind that the law of averages shows that the U.S. stock market (S&P 500) earns about 10% a year. But, investors cannot reasonably expect to gain 10% a year on average unless they remain in the market for several years, or even for a couple decades. While the S&P 500, which is an index of the 500 largest companies in the United States, has a 70-year record of going up about 10% a year, it can surge and drop dramatically within a short time.[25, 26]

The U.S. stock market rose more than 100% from 1995 to 1999; however, it then dropped 50% in the next few years. The NASDAQ, which included these dotcom technology companies mentioned earlier, had an even more drastic drop, down about 75% in two years. Recently the DOW dropped 777 points in just one day—it is down almost 40% from one year ago. However, it also recently surged 936 points in one day, a gain of 11%.

What do many investors do when they see their $100,000 investment shrink to $25,000? Rather than being patient by holding on to the stock and letting the law of averages increase the value of their investment, they panic and sell the stock. In effect, they buy high at $20 a stock and sell low at $5 a stock. This is their first mistake. Unfortunately, they tend to make another mistake soon after this that makes matters worse. Hoping to bounce back from their loss, they do not buy another such stock that has been devalued by a tumble. Instead, they buy a stock that is being aggressively advertised because it has gone up so dramatically in the last few months. In other words, they are again buying stock at inflated values. This is known as *performance chasing*. Buying stocks that have recently and dramatically surged sounds logical and intuitively correct. Selling stock that has recently and dramatically fallen also sounds logical and intuitively correct. Actually, both of these actions are usually illogical and unwise. These behaviors virtually guarantee that investors buy high and sell low, the exact opposite of what investors should do. Those who do this are guilty of the fallacy of regression to the mean. Usually such investors are far less successful than those who are patient and let the law of averages operate. Overwhelming evidence supports this suggestion, yet many investors are simply unaware of how violating this simple principle often leads to poor results.

Doctors sometime become so exasperated with the reasoning of their patients that they occasionally simply surrender and prescribe what their patients demand. Frequently the patients are guilty of committing this fallacy of *regression to the mean,* as well as the *fallacy of a common cause*. For example, suppose a patient has a fever and believes that this fever is causing his headaches. The doctor attempts to explain that the fever and headache are probably caused by another variable, such as a viral infection. The patient begs the doctor for a prescription for antibiotics; however, the doctor tries to explain that antibiotics are effective only in fighting bacterial infections; they are not effective for viral infections. Despite the logic of the physician, the patient demands antibiotics. A few days later, the now-healthy and gloating patient, calls the doctor to report that the antibiotics *worked.* This supposedly proves that the doctor was wrong and the patient was right. This is a faulty conclusion.

In all probability, the *fallacy of regression,* not the antibiotic, explains why the patient returned to a healthy condition. Quite simply, many people see their physicians at the height of their symptoms. It is at this point that their health is on a streak of abnormality—normally they are healthy, but now they have had several days of increasingly severe symptoms. It is also at this point that their symptoms typically become better or worse. For the vast majority of relatively minor ailments, their symptoms will get better without any treatment. That is, their bodies are returning to their previously normal and healthy states. This is not to suggest that patients should avoid seeing their doctors for their ailments. Although many medical problems will resolve on their own, some will become far more serious if they are not treated early. Even for minor ailments, initial treatment can speed up the recovery time.

Similarly, it is not a coincidence that as soon as the media identifies the oldest person in the world, she usually dies within a couple weeks. A woman who is 116, for example, has lived nearly 30 years beyond the norm. Regardless of whether the media identified her or not, chances are she would not live much longer.

Recall that the previous fallacy, the *Texas sharpshooter,* tells us that significant fluctuations, or exceptional short-term streaks of high or low performance are to be expected over the long haul. In contrast, this *fallacy of regression to the mean* tells us that these streaks will invariably end and that normality will return. Rather than realizing that the event (loss of investment or loss of a game) was caused by a simple regression to the mean, they tend to blame some other variable, such as a curse, bad luck, getting married, getting divorced, a stock broker, or even a magazine cover.

Fallacy

of

Gambler

Honey, grab that slot machine over there—the woman who just left played it for more than 20 minutes, and it didn't pay anything. It's bound to start paying out soon!

The *fallacy of the gambler (gambler's fallacy)* is a faulty way of thinking, in which one believes that a specific random-occurring event will be influenced by events that recently occurred.

This particular fallacy is committed by tens of thousands of people everyday in Las Vegas. Although many people know that the statistical probability of losing money in Vegas is greater than making money, they are willing to gamble because it is entertaining and exciting. However, there are many gamblers who have mistaken notions about winning.

Grab that slot machine—it hasn't paid off in a long time—so it is due to pay off soon. This statement is made by thousands of people every day in casinos across the country. They are committing the *gambler's fallacy,* a type of reasoning that is common, faulty, and costly.

Such gamblers fail to understand the most basic principles of probability and randomness. A slot machine that has not paid off for a long time is *not* more likely to suddenly start paying off. Nor is a machine that just paid a jackpot any more likely to stop paying out. Each outcome is random. Every machine has a computer chip that is essentially a *random-number generator.* The random number that is created is completely unaffected by the previous number that it created.

The random number generator in a slot machine functions in a way that is very similar to flipping a coin. When one flips a coin, it has a 50% chance of landing heads up. Suppose a fair coin has been tossed nine times and heads came up every time. What is the probability that heads will come up the next time, which is the tenth coin toss? Before answering this question, we must remember that the coin has no *memory.* The coin obviously has no ability to remember whether its last flip produced heads or tails. In other words, what happens next is not influenced by what just occurred previously. After nine heads, the probability that heads will come up on the tenth throw is exactly the same as it was on the first throw—50%.

What is the probability of throwing ten heads in a row? The chance is just one in 1,024. Note that this is the probability *before* tossing the coin. But if a coin has *already* come up heads for nine times in a row, the chance that it will come up heads on the tenth throw is still 50%. The previous coin tosses have nothing to do with the tenth coin toss.

What is the probability of picking the king of clubs from a normal deck of playing cards? Pirie Madsen, in his book, *How to Win Every Argument* (2006), reminds us that it is exactly the same as picking any other card—1 in 52.[27] If one picks the king of clubs, replaces it, and shuffles the cards, what is the chance of picking this card again? It is still 1 in 52. What is the probability of being dealt a particular royal flush in poker? It is exactly the same as for any other combination of five cards.[28]

This applies to slot machines, the roulette wheel, and a variety of other games of random chance. The odds of a slot machine paying off are not influenced by whether it recently paid off or not. If a machine just paid off a thousand dollars, the probability that it will pay off a thousand dollars on the next pull—or on any other pull is exactly the same. Every time the handle is pulled, the machine has the same probability of paying off. No trace of what just happened influences the next pull.

Just as a side note, some of the information in the previous paragraph is not 100% technically correct because it is not actually the *pull* itself that determines whether a slot machine will pay out. For most slot machines, the *timing* of the pull is what determines the payout. The computer chip generates hundreds of numbers between pulls, and the exact time at which the handle is pulled determines the payout.[29] Suppose you spend hours and hours at a machine, and then suddenly decide to quit. Immediately after you leave, you see someone sit down at *your* machine, pull the handle, and win $100,000. You should not be too upset by this because it was not the next pull that determined the payout, but the precise timing of that pull. The chance that you would have pulled the handle at precisely the same time the lucky winner did is extremely remote.

The gambler's fallacy is one that entices people to make foolish decisions on a regular basis. It is this lack of understanding that extracts millions of dollars from people each year. The millions, of course, are put in the hands of the casino owners, people who usually have an extremely keen understanding of probability and human behavior.

Fallacy

of

Superstition

People tell me I am crazy to be so superstitious, but I think they are wrong. The number 13 is bad news. The only Apollo mission that went wrong was 13. You can't deny that.

The *fallacy of superstition* is a faulty way of thinking in which one assumes that some irrational, mystical, or supernatural force causes certain events.

The *fallacy of superstition* is actually a subset of the *post hoc fallacy*. With superstition, however, the faulty causes relate specifically to so-called paranormal, or magical, associations between good and bad luck and specific actions, items, or numbers.

Astrology is perhaps the definitive example of this fallacy. Despite the fact that millions of Americans believe that one's astrological sign causes specific personalities or actions, there is no credible evidence supporting such a relationship. Both the theory and practice of astrology have been thoroughly debunked. Those who genuinely believe in astrology, unfortunately, are prone to many simple thinking errors, emotional pressures, frauds, cons, and deceptions.
There are also millions of people who consult astrologers or psychics *just for fun.* There is probably little harm in this practice as long as it is done only occasionally and strictly for entertainment purposes. However for some people, just visiting a psychic for *fun* sometimes evolves into a genuine *slippery slope.* Consider Arthur C. Clarke's quote: *I don't believe in astrology; I'm a Sagittarius, and we're skeptical.*[30] Although this is intended to be simply a paradoxical quote, it suggests that sometimes our behavior doesn't match with what we say. For example, we may say we are not superstitious, but often our actual words and behaviors contradict our claim. Although people may deny it, those who dabble in astrology sometimes begin to see their astrologers or psychics as genuine and legitimate advisors, confidants, or counselors.

Often when people have had some emotional or relational trauma, they rely increasingly on psychics. So-called *psychic hotlines* entice people to seek relationship advice with a simple phone call—and a credit card. As will be mentioned in another chapter, when *people become emotional, they often become stupid.* In this emotional state, they are much more susceptible to persuasion and manipulation. Psychics can often create very strong bonds of trust and credibility by performing impressive illusions and delusions through simple trickery or fraud. In so many of these illusions, people fail to engage their brains and instead let visual and psychological tricks fool their eyes and ears. Even when performers say their illusions are nothing more than tricks, some people are convinced that these performers have supernatural abilities.

Some people ask, *if psychics are not real, why do some police departments rely on them?* There are several simple reasons. First, there are some police chiefs who *believe* in psychic abilities; consequently they hire psychics. Second, if the masses of people in a community believe in the ability of psychics to solve crimes, some chiefs are simply responding to public political pressure. That is, they will hire a psychic just to satisfy the public, especially a public that occasionally works itself into an emotional frenzy over unsolved brutal crimes. Third, if a defendant believes the psychic has some supernatural power to identify the guilty, perhaps he is a bit more likely to confess. Finally, a psychic might have a slightly different perspective on human behavior and motivation; he may see some things that a detective doesn't necessarily see. However, none of these abilities are *psychic* abilities. They are simply heightened awareness or psychological abilities. A trained psychologist, for example, looks for cues of anxiety, discomfort, or mental disorders by analyzing the movement of one's body, face, and eyes, or by listening to the stress or pitch of one's voice. A *psychic* also employs many of the same psychological skills.

Perhaps the easiest way to force people to confront astrological claims is by looking at the *Barnum effect*. This is a term used to describe a common perceptual or thinking error in which people take general statements and actually believe them to be specific insightful statements about themselves.[31] Astrologers and psychics are masters of using this effect. Suppose we conduct an experiment by asking people to provide specific information, such as their date and location of birth. Each person is then told a professional astrologer has been assigned to his or her case, and will use this information to produce a specific analysis of the individual's personality. In reality however, the same generic analysis is given to each of the individuals.

> *You have a strong need for others to like you and for them to admire you. At times you are extroverted, affable, and sociable, while at other times you are introverted, wary, and reserved. You have a great deal of unused energy that you have not turned to your advantage. While you have some personality weaknesses, you are generally able to compensate for them. You prefer a certain amount of change and variety and come dissatisfied*

> *when hemmed in by restrictions and limitations. You pride yourself on being an independent thinker and do not accept other opinions without satisfactory proof. You have a tendency to be critical of yourself. Some of your aspirations tend to be pretty unrealistic.*[32]

After reading the analysis, the typical individual is quite impressed with its apparent accuracy. That is, the vast majority of people, when given general information can think of specific situations in their lives that seem to make the general information seem highly accurate and insightful. For example, when individuals call a *psychic hotline,* the psychic will usually try to develop trust and credibility with a few standard lines: *I am sensing great frustration in your life . . . you have to make a major decision in the near future . . . you have concerns about your relationship.* Although a rational person can see that these statements are so generic that they apply to virtually everyone, an emotionally distressed individual believes the psychic is a confidant with superhuman abilities.

This Barnum effect is similar to the *cold readings* that psychics, tarot readers, or palm readers sometime use. Cold readings typically begin by providing generic statements to clients and then watching their reactions. Often those who seek out psychics for advice are already psychologically vulnerable to suggestion or show strong emotional or psychological cues as to how they are feeling. These reactions help guide the psychic to make more specific statements or ask questions. Often clients are provided with very specific statements; however, there are a large number of extremely specific statements that are common to the general population.

For example, suppose an engaging and comforting palm reader studies a woman's hand then looks deeply into her eyes and whispers, *there is something that is causing you shame. You have hurt someone you love dearly, and this troubles you.* If the woman then looks down, the psychic may take this as a sign of guilt and continue with a line of questioning exploring these feelings. As implied before, such psychics may sincerely believe they have supernatural abilities; however, such readings often simply use nonverbal and psychological cues to tell people about themselves. Any good salesperson, counselor, or psychologist employs similar strategies to gain insights about what people are thinking and feeling.

There are hundreds of techniques that can be used to provide statements that seem amazingly specific and insightful. See Lynn Kelly's *Skeptics Guide to the Paranormal* (http://www.scribd.com/doc/14338371/the-skeptics-guide-to-the-paranormal) for an introduction to how this is done. Also, consult Ray Hyman's article about classic investigations of the paranormal, called *Cold Reading: How to*

Convince Strangers That You Know All About Them. More details can be found at: *(http://www.skepdic.com/Hyman_cold_reading.htm).*

Further readings can be found in his book, *The Elusive Quarry: A Scientific Appraisal of Psychical Research.* More details can be found on the following Website: *www.amazon.com/Elusive-Quarry-Scientific-Appraisal-Psychical/dp/0879755040.*

In reference to the example at the beginning of this entry about Apollo 13, in the United States a common superstition is the belief that the number 13 is an unlucky number. So common is this belief that many hotels do not have a floor labeled number 13. Breaking a mirror, seeing a black cat, spilling salt, walking under a ladder, opening an umbrella indoors, and stepping on a crack are associated with bad luck. Four-leaf clover, a rabbit's foot, crossed fingers, found pennies, and shooting stars are supposedly related to good luck.

Common perceptual errors often reinforce such superstitions. In fact, many social psychology textbooks explain a host of such flaws in our thinking and perceiving. It doesn't take much for some people to develop very strong superstitions. Many athletes, for example, will admit that they go through rituals or carry certain items with them in hopes of winning the big game. The belief that a penny, shirt, bracelet, necklace or other item causes good luck is emotionally etched in one's brain as soon as a win occurs. People tend to remember the *hits* or the times in which such an item seemed to create good luck. However, people tend to forget the *misses,* the times when the so-called good luck charm didn't work.

Believing a witch doctor can put a curse on someone is superstitious as well. However, if one genuinely believes that others have such powers, the belief itself can create a self-fulfilling prophecy. In other words, if you believe you are cursed, you will act in ways that produce undesirable effects. If you believe you will win, your confidence will produce behavior that will, in fact, increase the probability of winning. Many patients in medical studies report feeling less depressed or less anxious after being told that the pills they have been taking were antidepressants or anti-anxiety medications, despite the fact that the pills were nothing more than sugar pills (placebos). This *placebo effect* is well documented in medical and psychological research.

The *placebo effect,* or the power of positive thinking, influences certain events. However, it is the belief or delusion that influences further events. Believing that an actual item, whether it is a penny, shirt, or rabbit's foot, directly causes a particular outcome is a fallacy.

In speaking about psychic abilities, astrology, extrasensory perception, mental telepathy, and other so-called supernatural phenomena, the American Psychological Association reviewed more than 130 years of research on such matters and found no scientific justification for such claims.[33] Despite this fact, 50% of Americans still believe in ESP.[34]

Similarly, the James Randi Educational Foundation, an organization devoted to determining whether anyone has such psychic or supernatural abilities offers a *one million dollar prize* to anyone who can demonstrate such abilities under scientific conditions.[35] Specific information about the challenge can be found at www.randi.org.[36] Over the years and throughout the world, hundreds of famous *psychics* and others with so-called *supernatural abilities* have accepted the challenge. It should also be noted that *both* the Randi Foundation and the psychic being evaluated agree to the conditions of the testing in advance. Of these individuals, *not a single one* could demonstrate any such abilities.[37] This is not to suggest that all these individuals are cons or charlatans. Because all of these applicants agreed to the testing conditions prior to their performance, one could logically assume that many, if not most, genuinely believed that they had such powers.

Occasionally there are a few isolated studies around the world that claim such abilities exist; however, upon analysis, the evidence is so poor that it does not begin to approach the standards that high-quality science demands. Carl Sagan, one of the world's most well-known astronomers, authors, and professors, summarized the consensus of the scientific community when he said simply, *extraordinary claims require extraordinary evidence.*[38] That is, people who claim to have such extraordinary abilities are obligated to provide clear and convincing evidence of their claims. The extraordinary claims made by psychics and astrologers have not been supported by extraordinary evidence. In fact, as stated previously, such claims have not been supported even by ordinary evidence.

Believing that seeing a shooting star caused one to win $100 from a lottery ticket is perhaps a minor and seemingly harmless superstition. However, developing an obsessive-compulsive disorder of never stepping on a crack in the sidewalk because it will supposedly cause some disastrous event is far more serious. The same type of reasoning often leads people to believe in the *supernatural powers* of so-called faith healers.

It is certainly true that the good feelings or positive thoughts that many people derive from their religious beliefs should not be minimized. Having a positive attitude and faith can clearly affect one's health. But, rationality must be maintained. Although conventional medicine is highly effective for millions of people with hundreds of conditions, it cannot relieve all pain. If conventional medical treatments have not worked, one would certainly be justified in seeking other experimental or alternative forms of care. But, there are people who reject the rationality of evidence-based conventional medicine initially, and put their medical treatment in the hands of so-called healers who rely on chants, rituals, mysticism, and superstitious practices. Such beliefs can lead to death.

Chapter 5

Fallacies of Authority

Ad Hominem (Genetic)

Ad Hominem (Abusive)

Ad Hominem (Circumstantial)

Guilt by Association

Tu Quoque

Poisoning the Well

Ad Vercundiam

Chapter 5

Fallacies of Authority

There is an ancient Chinese proverb that describes four kinds of people.[1]

Those who don't know, but don't know that they don't know.
Those who don't know, but know that they don't know.
Those who know, but don't know that they know.
Those who know, but know that they know.

These four categories of people reflect different levels of competence or authority. This chapter illuminates some of the ways in which one's credibility or authority is used incorrectly in defending or rejecting a particular argument.

Those in the first category believe that they are authorities; however, those who listen to them quickly realize that is not the case. To make matters worse, they tend to be extremely confident in their conclusions simply because they do not even pause to consider whether their information is correct, or whether their reasoning is valid. They are the *know-it-alls,* but they are totally unaware that others perceive them in such a negative way. If such people are in positions of authority, like dictatorial bosses, we often feel compelled to agree, so we just nod and smile. If we do question their conclusions, they often become defensive and counteract by accusing us of being incompetent. Ironically, the brain living inside the head of such a person is actually quite happy—it never questions or doubts itself. Of course the owner of this brain is delusional. Indeed, as widely quoted, *ignorance is bliss.*

The second group of people openly acknowledges their lack of authority on particular issues. Because they have a good understanding of what they do not know, they are open to learning. They also seek out others who have authority in areas that they do not and then use that evidence and those reasons to make better decisions.

The third group of people are seen by others as having more authority or credibility than they think they have. Others come to them for wisdom and insight, but they fail to realize or appreciate their own talents.

Individuals in the fourth group of people are authorities, a fact known to them and known to the people around them. They are knowledgeable, and they use this knowledge to help themselves and others who seek their advice. But, they are not arrogant or overconfident. Although they may have authority in many areas, they too share some of the same characteristics of those in the second category. No one can be an expert in all things, a fact that they never forget. So, they too seek out authorities when they need to. They have authority because they take great care to ensure that their information is accurate and that their reasoning is valid. But, they also have authority on their own authority—they know when they have authority and when they do not.

Keep these four types of people in mind as you travel the roads of this chapter. What types of fallacies do some of these individuals commit? How can authority be misused? Is it fallacious to believe a claim made by a famous actor on a particular medical treatment, for example? Should one accept the authority of one's dentist, for instance, on issues of investing?

Authority is an element in logic and persuasion that is of utmost importance. As suggested in Chapter One, for thousands of years, the study of persuasion has focused on three basic elements: emotions, logic, and credibility. The Latin terms, which are still used sometimes today, are *pathos, logos,* and *ethos.* Evidence indicates that persuasion is more likely to occur if one can activate the emotions of the audience (pathos), appeal to their sense of logic (logos), and finally, demonstrate speaker credibility or authority (ethos).

We will look at a number of fallacies related to pathos, or emotions, later in the book, but first we turn our attention to ethos, the credibility or authority of the one making an argument. Considering the source or speaker is sometimes appropriate in determining whether to accept evidence in support of a claim; however, in many cases it is completely irrelevant. For example, rejecting the claim that $2 + 2 = 4$ because a convicted murderer made this statement is a perfect example of this fallacy.

However, it is not a fallacy to question the statement, *I have never given false testimony,* from a twice-convicted perjurer. But, rejecting an argument based solely on a speaker's questionable character, competence, or motives, is not appropriate or relevant to many arguments. Again, it really doesn't matter whether the claim that 2 + 2 = 4 comes from a mathematician or a convict. The source of the argument does not affect the truth of the conclusion.

In speaking about the authority of the fourth person above, from a logical and technical point of view, we should NOT accept his statements as *the truth* simply because *he is an authority.* Why not? Whom can we trust, you might ask, if we can't trust an authority? He could, of course, be biased in providing his statements, or perhaps even attempt to deceive or give statements he knows to be false because of some other ulterior motive—perhaps blackmail or financial reward. Or, perhaps, his so-called authority is nothing but a deceptive facade.

However, there is even a more fundamental reason, a reason that serves as the foundation for valid thinking, as to why we don't normally accept the truth of a conclusion because *he said so.* To be more precise, statements from authorities can appropriately influence the conclusion we accept because we know the authority can provide *evidence* and *valid reasoning* to support his assertions. We don't, for example, directly accept the truth of evolution simply because *Charles Darwin said so.* We accept it because Darwin has provided extensive evidence and scientific reasoning that supports what he says.

If a physician, for example, suggests a particular treatment for your medical condition, a critical thinker asks *why.* As an authority, the doctor should then be able to provide evidence and reasons for recommending that particular treatment.

To be slightly less theoretical and a bit more realistic, it would be impossible to discuss all of the evidence and reasoning that a physician, financial advisor, educator, or other authority might consider in making a recommendation to you. In other words, we will not suggest that your critical thinking is deficient if you sometimes *trust* a conclusion provided to you by a legitimate authority. That trust, of course, is predicated on evidence and reasoning, both of which could be provided if asked for.

Despite the latitude discussed in the previous paragraph, we need to be especially careful of assertions made by a person who claims to be an authority. As mentioned previously, one goal of education is to provide *answers to questions.* However, the inverse is even more important—to *question the answers.* Knowing, for example, how authority is misused in both rejecting and accepting answers is the focus of this chapter.

Fallacy

of

Ad Hominem (Genetic)

I have to vote against the bill introduced in Congress extending unemployment benefits for another six weeks. Why? Well, the original idea for providing money to those who are unemployed actually stems from communism.

The *fallacy of ad hominem (genetic)* is a faulty way of thinking, in which one rejects a claim, not based on its merits, but on the *origin* of the idea.

In all probability, most students have heard of the term, *ad hominem.* It is a widely used Latin term that means *against the person.*[2] It is a broad term used to describe various faulty ways of thinking in which one rejects a claim based upon the authority of the source, and not on the argument itself. Put another way, the claim is rejected, not for *what* it says, but because of *where* the claim came from. So, the key in identifying these fallacies is to determine whether individuals are arguing primarily *against the person* or against the *reasons* for the argument.

There are actually several different types of *ad hominem* fallacies, and much of this chapter is specifically devoted to them. Some will have more specific and unique names, although three of them, including this one, will include a term in parentheses to denote how it is different from the other *ad hominem* fallacies.

Genetic, as used in logic, refers to *genesis,* a term that relates to the birth or origin of an idea. *Genesis* is one of several books in both the Bible and the Torah that deals with *creationism,* or the origin of life.

In the example above, the suggestion is that supporting the unemployment bill would, in effect, be supporting communism. In other words, no logical discussion of the actual bill occurs—it is dismissed because one rejects communism, the origin of the idea. This fallacy is persuasive and effective because some people will transfer any undesirable aspect of an idea's origin to the idea itself. Nearly any candidate or idea that can be associated with communism can be defeated, no matter how trivial or coincidental the connection might be. This is true because it is etched in the consciousness of most Americans that communism is a deeply rooted evil.

Because some people identify themselves so strongly with a political party, they will often refuse to even consider a bill because it was introduced, or originated from, the opposing political party. In fact, so strong is this political disposition that some will defend their rejection of the bill by smugly declaring, *just look at the source.* Their smugness reflects the characteristics of that person who *doesn't know, but doesn't know that he doesn't know.* Rejecting an idea outright based specifically on where it was born is illogical. The merits of an idea can be, and usually should be, evaluated apart from where it originated. Again, simply saying the idea is deeply rooted in a *liberal* or *conservative* philosophy is not grounds for its immediate dismissal.

In fact, many people react emotionally to these terms, usually negatively; however, they have little understanding of what the terms mean or imply. When quizzed on where most *liberals* or *conservatives* stand on a wide variety of issues, many voters are dumbfounded. And when they do answer, their answers are often extremely superficial and simplistic. They just *know* that fill-in-the-blank (liberals or conservatives) are *bad*. They know they are *bad* not because of proper evidence or valid reasoning, but because someone *told them* they were bad. To fight this ignorance and irrationality, and to improve the chances that a bill will pass both Congress and public opinion, both a Democrat and a Republican often cosponsor bills.

Religious philosophies can also be exploited for this purpose. It is not uncommon, for example, to initially find support for a new library or other such project in a community. However, one will frequently find that this support suddenly turns to opposition when it is revealed that the initial proposal came from a group of Atheists. In many cases, whether the original idea came from Atheists, Christians, or Muslims is irrelevant to the merits of the idea.

Obviously it would be naïve to think that the source is *always* completely irrelevant to how people *perceive* the merits of a claim. Suppose, for example, that a public school is determining whether to accept new biology textbooks that are being donated from a particular organization. If these textbooks are being donated by a far-right religious organization, it is appropriate to question whether the organization has any ulterior motivation for the donation. Do the textbooks replace the scientific principles of evolution with theological doctrines of creationism? If so, then the donation should be rejected.

But, even in this case, the origin of the idea, or group, should only make one more careful or suspicious in investigating the truth of the evidence for the claim. It should not serve as the definitive reason as to why the books were rejected. The contents of the books can be evaluated completely apart from where the books originally came from.

Several other *ad hominem* fallacies will be examined later in the chapter. They all have one common theme—rejection based on the *source* of the claim, not on the *substance* of the claim. Again, this genetic version is a bit different from the others in that it focuses specifically on the *origin* of the claim.

Fallacy

of

Ad Hominem (Abusive)

> *The reporter's claim that the war in Iraq, as of March 2008, has already cost more than ten times the White House's original 2003 estimate is ridiculous. He is wrong. You can't believe him. He is such a pathetic person—fat, bald, ugly, and stupid. He is such a loser that all he does is spend his life criticizing others. He is just an unpatriotic traitor, who wants to end the war. He will say anything just to get his face on TV.*

The *fallacy of ad hominem (abusive)* is a faulty way of thinking, in which one personally attacks or insults the person making the claim, rather than discussing the claim itself, in an attempt to give the impression of discrediting the claim.

In the example above, is the claim about the war's cost true? Even the Pentagon and the Congressional Budget Office have verified the accuracy of the claim.[3]

In fact, well-respected authorities are now providing evidence and reasons to support their estimates that the cost of the entire war will be somewhere between two and three trillion dollars—almost *50 times* as much as the original estimate—or approximately $8,000 dollars for every single person in the United States.[4] Joseph E. Stiglitz, a Nobel Prize–winning economist, calculated the cost to be higher—at more than three trillion dollars.[5] Even conservative estimates from the Congressional Budget Office say a two-trillion dollar estimate is realistic.[6]

Do note that the reporter makes no direct claims as to whether the war is worth its financial costs or not. Nor does the reporter explicitly state he is against the war, or that the war should end. Nor does he suggest that the war was launched due to the mistaken belief that Iraq had weapons of mass destruction. He is simply stating that the original estimates turned out to be off by at least a factor of ten—or 1,000%.

The claim made by the reporter can be easily substantiated by numerous independent sources with concrete evidence. However, even if the claim was wrong, the individual who verbally attacked the reporter committed a fallacy. Rather than dispute the accuracy of the claim, he tried to give the impression that the claim was untrue simply because the person reporting it had personal characteristics that he despised. He used a tactic that is very familiar to most attorneys—if one cannot dispute the claim, attack the person providing the testimony. Of course, many effective attorneys are a bit more restrained and subtle in their personal attacks.

He is such a pathetic person . . . fat, bald, ugly, stupid . . . loser . . . unpatriotic traitor—clearly these statements are insulting and abusive. Most of the insults are subjective opinions as well. And even if the reporter is generally perceived to have these personal traits, they are irrelevant to the claim he makes.

The abusive individual was trying to get people to reject the reporter himself, and therefore transfer that rejection to the claim itself. Unfortunately, the masses of people often fall for this manipulation and often do exactly that—reject a claim because they are persuaded to dislike the person making the claim. This is one reason why politicians, attorneys, and persuaders in general, employ this tactic so often. In fact, millions of voters who are totally uninformed about a candidate's position often base their vote on nothing more than a comparatively irrelevant personal characteristic.

Negative advertising from the opposing candidate is often highly effective in creating not only dislike, but also resentment and hatred. Even when voters agree with a candidate's stance on all of the major issues, they can often be persuaded to vote against such a person if the opposing campaign can exploit some trait that voters dislike.

Abusing or insulting a candidate by calling him or her an *idiot, an elitist, a moron, a womanizer, a man-hater, a racist, arrogant, unpatriotic,* or *sexist* can activate powerful and negative emotions that often overpower voters' rational reasons for voting for a candidate. Using these words can serve as powerful insults; however, even describing a candidate as an *intellectual* or *privileged* can often trigger negative feelings in those voters who want to elect *a regular guy.*

This is not to suggest that all attacks on a person's credibility are inappropriate. The ad hominem (abusive) fallacy occurs only when the person is being personally insulted, verbally abused, or attacked about some irrelevant aspect of his character or traits. Again, in determining the truth of the claim about the cost of the Iraq war, attacking the personal characteristics of the reporter is completely irrelevant to whether the report is true. In contrast, an attorney is justified in attacking some aspect of a person's character if it is relevant to determining whether the person is generally honest and truthful. Suggesting that someone is a habitual liar, and that his testimony should be questioned is clearly personal; however, it is not a fallacy if his habit of lying can be established by evidence.

The determining factor, of course, is whether the inquiry about some aspect of a person's personality has a bearing on the statements he is making. As just mentioned, attacking the truthfulness of a witness is sometimes justified. In contrast, suggesting that person is lying because she is *ugly and stupid* would be a fallacy.

In review, the *ad hominem (abusive)* fallacy occurs when one uses an irrelevant and personal attack on a person's character as a ploy to convince others that the argument itself should be rejected or accepted. This fallacy is sometimes called *character assassination,* the *smear campaign,* or *mudslinging.* The previous *genetic* fallacy focused on rejecting a claim based on its *origin.* This *abusive* version also tries to get people to reject an argument, but it does so by attacking the person's character or appearance, while being insulting or abusive.

Fallacy

of

Ad Hominem (Circumstantial)

> *Let me remind the voters that my opponent wants to repeal the capital gains tax because he says it would increase unemployment. What he doesn't tell you is that he would personally save over a million dollars in taxes if that happened. Of course, anyone in his shoes would say the same thing. Obviously, his position is wrong.*

The *fallacy of ad hominem (circumstantial)* is a faulty way of thinking, in which one rejects an argument simply because the one making the claim would benefit if it were true.

As the name implies, the circumstance or situation of the one advocating a claim is used as the reason why it should be rejected. Does the fact that the advocate would personally gain if the claim were true actually affect its genuine truth? Logically, the simple answer is that it does not. Although strong circumstantial evidence of personal gain is not to be ignored, it does not, by itself, provide definitive proof that the claim is not true. This is one reason why in criminal cases where the standard of proof is *beyond a reasonable doubt,* circumstantial evidence is rarely enough for a conviction.

Suppose, for example, that a well-known nutritional researcher completes a study of two popular diets, and concludes that the individuals on one diet lost significantly more weight than did those on the other diet. Then, it is revealed that the founder of one of the diets funded the study. This fact does not make the finding false. Note that so far we do not yet know whether the sponsor's diet or the competing diet was found to be most effective. If the competitor's diet turned out to be the most effective, many people would accept the study as being legitimate. After all, the sponsor of the study did not benefit from the outcome of the study. On the other hand, if the results supported the sponsor's diet, many people take that as definitive proof that the finding is untrue. Such a conclusion violates the principles of logic.

Using this as the sole basis for rejecting the truth of the study is not logical; however, it is certainly reasonable to investigate the evidence more thoroughly than we might otherwise. At minimum, it can create the appearance of impropriety, even if none exists.

In the example about the effects of a capital gains tax, the claim that it leads to unemployment can be judged on its own merits. Whether it helps or hurts the one making the claim is not logically relevant to whether it is true. Again, though the conclusion is not logically relevant, many people are psychologically predisposed to automatically reject claims that appear to be self-serving. In reality, the fact is that many, if not most claims are initiated from those who would benefit from their acceptance.

It might be appropriate to question the ethics or expertise of the one making a claim that seems to be self-serving. But personal benefit alone is not reason enough to reject the claim. If this were the case, most arguments would be instantly dismissed or rejected with *of course, he is for/against ____________ because ______________*. *Of course he is going to argue that birth control is wrong—he is a Catholic priest. Of course, he is going to argue that healthcare should be universal—he is a doctor. Of course he is going to say that the drug is safe—he represents a drug company. Of course he is going to say that education requires more funding—he is a Democrat. Of course he wants to spend more on the military—he is a Republican.* These claims can be evaluated apart from the one making the claim.

This is not to suggest that a person's self-interest is always completely irrelevant to judging the overall claim. In a court of law, establishing a motive for a crime is often a key component of a persuasive case. The fact that a wife takes out a million-dollar life insurance policy on her husband just weeks before he is found dead is certainly suspicious. But this fact alone is relatively weak evidence; a jury returning a murder conviction for the wife solely on this evidence probably is guilty of committing an *ad hominem (circumstantial) fallacy.* Establishing a motive is just one part of developing a good case.

It could be argued that almost all human behavior has some self-serving component. Giving to charity, for example, helps others, but doing so often makes you feel better about yourself. Perhaps it creates a sense of appreciation, relieves some sense of guilt or obligation, or promotes causes you value. Perhaps it is simply done to get a tax deduction.

Most United States senators, for example, initiate bills or arguments that are self-serving, at least in the sense that the bills would be beneficial to the state that those senators are representing. And politicians know that bringing federal dollars to their state often helps get them reelected. But, the self-serving motivation behind sponsoring the bill is not reason enough to reject the bill.

Many attorneys, for example, advance arguments in personal injury cases because doing so will help them win a case—and a share of the settlement. Logically, the fact that they are essentially being paid to initiate arguments does not make the claims any stronger or weaker. Whether the attorney is working a *pro bono* case (work without pay), or being paid $500 an hour, the argument does not draw strength from either fact.

In review, the *ad hominem (circumstantial) fallacy* tells us it is a fallacy to instantly reject an argument just because the sponsor's circumstance is such that he would personally benefit. Like the other *ad hominem* fallacies, it incorrectly suggests that the source, rather than support, is the dominant factor in determining an argument's strength.

Fallacy

of

Guilt by Association

I see that presidential candidate, Hillary Clinton, wants universal healthcare for every citizen in the United States.[7] *You know, Hitler was also in favor of universal healthcare.*

The *fallacy of guilt by association* is a faulty way of thinking, in which one rejects a claim, not based on its merits, but by associating the claim with a person or notion that people reject. It falls under the broad umbrella of *ad hominem* fallacies.

In the example above, associating Hitler with universal healthcare is a blatant example of this fallacy—and most people would see that this is nothing more than a rhetorical gimmick. It is an obvious attempt to encourage people to transfer their hatred of Hitler to the claim about healthcare.

In fact, this fallacy is sometimes called the *Hitler Card* or *Nazi Card*. It is a devious, but highly effective form of persuasion. Any individual who is unable to defeat a claim on the basis of evidence and reasoning can always resort to exploiting the hatred that people have towards Hitler. This hatred, although not logically relevant to the claim, can be psychologically persuasive. People will go to great lengths to avoid even the appearance of having any relationship, no matter how irrelevant, to Hitler or similar despicable leaders.

Some people have even gone so far as trying to link the vegetarian diet to Hitler. After all, they claim, Hitler was a vegetarian too. This is as far-fetched as saying exercise should be avoided because Hitler was an advocate of exercise.

Again, at a logical level, it is clear that these associations are completely irrelevant to whether one accepts or rejects a claim. But, psychologically, people tend to reject ideas if the ideas are somehow associated with people they dislike. In order for people to keep balance or consistency in their beliefs or behaviors, they often reject ideas or people that conflict with deeply engrained beliefs or ideas.

As an analogy, a shopper might find a new blouse that she thinks is especially attractive—in its own right it is rather appealing. However, upon taking it home, she realizes it clashes with her entire wardrobe. Consequently, she rejects the new blouse and returns it to the store. Psychologists call the idea of having conflict between one's ideas *cognitive dissonance.* Of course, *cognitive* refers to thinking or thoughts, and *dissonance* means discord or disagreement. This internal dissonance is rather uncomfortable, and it often creates anxiety. The easiest way to relieve this anxiety is to simply reject the new idea.

Critical thinking acknowledges that consistency in ideas, beliefs, and reasons is the hallmark of logic, but it also acknowledges that, in a complicated world, it is sometimes necessary to entertain ideas that conflict, or that seem to conflict. In the case of the blouse, perhaps keeping it might encourage one to build another wardrobe, or to replace the wardrobe that is outdated. Or the blouse might encourage one to wonder whether a conflict really existed—is it possible, with a bit of creativity, that it can peacefully coexist in harmony with other pieces of clothing?

Similarly, critical thinkers know it is possible to *like the singer, but hate the song.* Voters have a difficult time with this issue—some voters buy everything one political party represents—hook, line, and sinker. Critical thinkers should have the ability to say that they generally support the ideals of a particular party; however, they should also have the courage to realize that some of the tunes or ideas presented by their party are not worthy of their support. A bad idea—regardless of whether it came from a Republican or a Democrat—is still a bad idea. Furthermore, some people believe that supporting *any* idea that comes from the other party is similar to *giving aid and comfort to the enemy.* Such partisan and simplistic thinking often inhibits creative problem solving.

Ever since 2001, a number of government officials have suggested a *link* or *association* between the 9-11 terrorists and *Saddam Hussein.* Despite the fact that no credible evidence has supported a relevant link, was the mere mention of a link enough evidence for some to convince themselves that a war against *Hussein* was a war against the *9-11 terrorists?* If so, some government officials committed the fallacy of *guilt by association.*

Ever since Senator John McCain became the Republican nominee for president in 2008, he has been linked to President Bush. Democratic leaders have emphasized this association because President Bush had very low public approval ratings, and these ratings, it was assumed, might help the democratic nominee in the presidential election. Was linking McCain to Bush a relevant link or was it *guilt by association?* Linking McCain's stance on continued military presence in Iraq to Bush's stance is justified because their positions are almost identical. On the other hand, suggesting a similar link between their policies on expanding federal funding for embryonic stem cell research would not be correct—Bush opposes such funding, and McCain favors it.[8] These two issues are clear-cut, but how strong and relevant is the McCain-Bush association? Is this a case of *guilt by association* or is the link justified by evidence? Voters who carefully analyze the positions of these two men will, in fact, find a number of relevant similarities; however, they will find some differences as well.

In review, relevant associations or links between individuals can sometimes be used in logical reasoning. The fallacy of *guilt by association* only occurs when irrelevant links are purported to be relevant links.

Fallacy

of

Tu Quoque

Seventeen-year-old daughter: *Mom, you are such a hypocrite! Every time I go to a party you make me promise not to drink or take drugs. You are constantly lecturing me about how easy it is to get addicted to drugs. Dad said you drank like a fish in college and smoked dope almost every weekend. You, and especially you, have no right to tell me what not to do when I go to parties. I can decide on my own whether drugs are addictive.*

The *fallacy of tu quoque* is a faulty way of thinking, in which one dismisses a claim on the grounds that the very person making the claim is guilty of the behavior she is criticizing. Because it tries to negate a claim by irrelevantly attacking the source, it too is one form of the *ad hominem* fallacy.

Tu quoque is a Latin term that roughly translates into *you too*, or *you're one too.*[9] At the most basic level, an individual is rejecting an argument, not based on the merits of the argument, but rather on the grounds that it comes from a person who appears to be (or is) a hypocrite.

In the example above, if the mother tried to keep her alcohol and drug problems a secret, it would certainly be fair to call her a hypocrite when she now is so adamantly opposed to them. On the other hand, if she admitted her past problems prior to her criticism, most people would not say she is being hypocritical. In any case, and regardless of whether she disclosed her past or not . . . and regardless of whether she is a hypocrite or not, her past is logically irrelevant to the merits of the claim. Although it is true that we psychologically tend to despise people who don't practice what they preach, this inconsistency is not a logical reason to reject the claim.

The old adage, *one who lives in a glass house shouldn't throw stones*, attempts to capture this idea. The biblical quote, *let he who is without sin cast the first stone* is another way of saying the same thing. And, although these statements have some intuitive appeal, they are not logical. Although they might be considered *logical* by the masses of people who are unfamiliar with the rules or principles of logic, they clearly violate the basic principles of valid reasoning.

As mentioned previously, the notions of beliefs and truths are often not the same thing. The masses of people might believe that someone is guilty; but this psychological belief does not necessarily translate into logical truth. Frequently, what the masses of people believe is simply not the truth. In the case of this fallacy, people who accept the idea that one is not guilty because the accusation came from a guilty person are violating logical principles. The guilt of one person does not directly support the innocence or guilt of another person. Evidence and reasoning, not the source of the evidence, are the gold standard of determining truth.

Often an individual who is caught cheating contends that he should not be punished because his friend also cheated. Logically this does nothing to diminish his guilt. Actually it just suggests that his friend should also be punished.

Again, although counterintuitive, whether a claim comes from a hypocrite or from a saint, the truth of the claim is not determined by its source, but by careful examination and reasoning. Because we often allow the source to influence our acceptance of a claim, sometimes efforts must be taken to guard against this mental error. For example, in order to prevent bias or initial prejudice of a scientific finding, scientists reviewing the case are typically not told whether the researcher is a world-renowned, Nobel prize–winning scholar or an obscure graduate student.

Similarly, a common practice at some colleges is to protect the identity of students who submit essays for a writing contest. Essays, with the names removed, are forwarded to professors who then serve as judges. This forces the professors to evaluate each paper apart from a student's characteristics. Whether the student is male or female, prompt or tardy, attractive or ugly, young or old, should not affect the merits of the paper itself. As discussed previously, the *confirmation bias* error occurs when we don't separate the source from the content. If we like the source, we tend to focus on the positive aspects of her or his work and ignore or minimize any negative aspects. If we are in a happy personal relationship, we tend to focus on the positives of our partner. Of course, the inverse is true as well.

The claim in the initial example above, which focuses on whether it is easy to get addicted to drugs, can be evaluated on its own merits and apart from the source itself. Although such a claim can be more powerful and psychologically persuasive if it comes from one who has personal experience, scientific and medical evidence of addiction can be judged apart from its source.

Though psychologically difficult to accept, a doctor who secretly smokes two packs of cigarettes a day has as much logical legitimacy in encouraging his patients to quit as does a nonsmoking doctor. The Atkins Diet, developed by Dr. Atkins, was widely criticized when it was learned that the doctor was not particularly thin at the time of his death.[10] This fact alone is virtually irrelevant to whether the diet was effective or not in helping most people lose weight. Long-term, large-scale studies using careful research methodology, evidence, and logical reasoning produce findings that are closer to the *truth* than simply citing a single case that may not be representative.

Fallacy of Poisoning the Well

Attorney to Media: *This afternoon you will hear a person testify about what he* ***claims*** *happened that day in the bar. But before you listen to his story, you should know that he drinks heavily and smokes weed every weekend. In fact, he once was arrested for public intoxication, spent the night in jail, and, when he woke up the next morning, he couldn't even remember what had happened. Obviously anything he says is nothing more than what he dreamed up.*

The *fallacy of poisoning the well* is a faulty way of thinking, in which one tries to discredit what an individual might say in the future by attacking his credibility *before* he even has a chance to say anything. This too is a type of *ad hominen* fallacy, because it tries to get people to reject a claim by pointing out that the one making the claim is objectionable in some way.

The key word in this description is the word *before.* That is, it is essentially a preemptive attack on anything the person might say. In contrast, *after* a witness provides damaging testimony, many people expect that his credibility will then be immediately attacked—simply because he said something damaging. That is, they believe the sole reason he is being attacked is because he said something that hurt the attorney's case, and that he would not be attacked if he had actually said something beneficial.

A slightly less conspicuous approach is to give the impression that it is not genuinely known what the witness is going to say; perhaps his testimony might aid the case, or it might damage the case. But, because we are trying to be fair, we are saying upfront, regardless of what he says, that he is not credible and should not be believed. Although it might seem like an honorable approach, it is often a deceptive practice simply because it is usually known in advance what the witness is going to say. Perhaps this tactic creates some goodwill or trust in the same way that a politician might occasionally refuse to accept a large donation, for example, because it comes from a dishonorable organization.

Both the abusive and circumstantial forms of *ad hominem,* like this fallacy, attack the credibility of a person making a claim rather than the claim itself. They are simply different kinds of attacks. The abusive form relies on name-calling, degrading insults, or offensive labels; the circumstantial form simply says that the circumstances of the witness might make him suspect or biased. *Poisoning the well* might actually include one or both of these approaches, but its effectiveness relies on the fact that the attack is done before any claim is even made.

In looking at the first example, testimony is obviously compromised if it comes from someone who drinks heavily and smokes marijuana. But, this fact alone doesn't automatically prove the testimony is completely untrue. In fact, unless it can be shown that the person was drunk or high *at that time,* substantial credibility can be retained.

Often the fallacy of poisoning the well is, in itself, based on unsubstantiated accusations and charges. For example, simply declaring that the witness was drunk or high, even if it was false, is enough to plant a seed of doubt in the minds of those who listen to what the witness says. It is an attempt to prejudice the jury—or to make them predisposed to question his credibility. Although most of us say we dismiss the supermarket tabloid stories as fiction, is it not true that sometimes these stories make us wonder whether there is some truth to what is reported? Often simple accusations, myths, and rumors spread like wildfire, especially given the popularity of the Internet, and they frequently settle into the consciousness of many people as *truths.*

One particularly interesting way that people poison the well is to use a technique known as *every schoolboy knows.* In an attempt to preemptively silence any students who might disagree, teachers many decades ago sometimes prefaced their comments with: *Of course, every schoolboy knows this is true.* This poisons the well because it kills any dissent before anyone can disagree. There are various wordings that have the same impact: *no intelligent person would think otherwise; only an idiot would disagree with this idea; every reasonable person believes, etc.*

Ironically, a technique known as the *big lie* can often be even more effective. It is a propaganda technique that Hitler defined as so *colossal* that no one would even think about saying it unless it was true.[11] During the summer of 2008, the presidential campaign was becoming more contentious, and already such accusations were being made. Some asserted that Senator John McCain knew about the terrorist attacks on 9-11 long before they occurred. Similarly, some asserted that Senator Barack Obama was secretly supportive of the terrorists. Such vile, mean-spirited accusations, without having a shred of evidence, can plant a seed of doubt in the minds of voters. The statements are so incredibility outrageous and fantastic that they make people wonder whether there are, in fact, kernels of the truth hidden somewhere. Supermarket tabloids work in the same way.

These statements are vastly different from the accusations that *McCain is too old to be president* or that *Obama doesn't have enough experience to be president*. Instead, these statements tap into the propaganda potential of the *big lie.* Again, the goal is to plant a seed of doubt in the minds of voters by *poisoning the well.* No matter how fantastic and unsubstantiated such claims are, they are sometimes enough to trigger bias, prejudice, contempt, and hatred of a candidate.

In speaking about the name of this fallacy, suppose that an advancing army sets up camp close to the next town they intend to attack. If a soldier sneaks into the town during the night and drops cyanide into the water supply, he has literally *poisoned the well.* The next morning those living in the town drink from the well and fall sick or die. Consequently, the army has destroyed the opposition before it had a chance to fight or do any damage. Whether in a physical battle or verbal argument, those who use this fallacy believe that the best defense is a strong offense.

In high-profile cases, especially those involving celebrities or politicians, trying to preemptively smear the credibility of a damaging witness is common practice. The *court of public opinion* often influences a politician's ability to get reelected or a celebrity's opportunity to star in a film. It is also far easier to make malicious attacks on one's character in the media than in the courtroom.

One reason poisoning the well is so commonly used is that it is often highly effective and difficult to suppress. Becoming more skeptical of all the information we consume can sometimes lessen its damage.

Fallacy

of

Ad Vericundiam

Actor Tom Cruise on NBC's *Today Show* to host Matt Lauer (June 24, 2005): *Psychiatry is a pseudoscience . . . there is no such thing as a chemical imbalance in the body . . . you don't know the history of psychiatry. I do.*[12]

The *fallacy of ad vericundiam* is a faulty way of thinking, in which one accepts the truth of a claim, despite the fact that it comes from someone who is not a genuine authority in that field. It is sometimes called *appeal to inappropriate authority* or *celebrity endorsement.*

This fallacy is the converse of the *ad hominem* fallacy, which is based on rejecting a claim because of its source. In contrast, this fallacy (*ad vericundiam*) occurs when people blindly accept the truth of a claim because it comes from someone who is popular or well liked. This is not to suggest that people intentionally ignore the source's lack of credibility and choose to believe what he says anyway. Frequently, the source is so likeable, popular, attractive, or dynamic that we forget to ask whether he is credible or has the necessary expertise to make claims in a particular field.

Although an expert is not needed to help determine whether 2 + 2 = 4, experts often must be relied upon to verify the legitimacy of a claim. Most issues are far more complex than simple arithmetic, and people must necessarily rely upon experts to provide evidence and reasoning so that others can determine whether a claim should be accepted.

Advertisers know that famous musicians, actors, or athletes can be used as tools to publicize their products. It is perfectly legitimate to have a famous athlete advertise or endorse a particular product, as long as the athlete is a genuine authority in that specialty or product. Tiger Woods, a famous and extremely accomplished golfer, has the expertise and authority to endorse a particular kind of club or golf ball. He does not however, have the genuine authority to recommend Buick automobiles over any other brand.

Britney Spears, a famous musician and performer, developed her own line of perfume—and uses her popularity to promote these products. Although Spears has little authority on such issues as federal fiscal and monetary policies or terrorism, her perfume commercials are not blatant examples of this fallacy, simply because she expresses only a subjective opinion about a subjective product. Saying, *I love the way it makes me feel,* for example, is a personal emotional statement and is not technically even an argument; it does not require a careful analysis of the evidence and reasoning to determine whether the claim is true.

If an advertiser is using a celebrity simply to gain the attention of the viewer so that a persuasive and logical claim about a product can be made, a fallacy is not technically committed. Appealing to emotions (sex-appeal, adventure, loyalty, embarrassment, fear, etc.) as a way of selling products is common, and advertisers know that simply making a positive connection between some product and a celebrity is an effective marketing tool. But, when we try to determine the truth of a claim through emotions, rather than using evidence and logical reasoning, fallacies begin to emerge. The inappropriate use of emotional appeals will be explored in another chapter.

Many explicit examples of *ad vericundiam* occur when elderly actors or celebrities make appeals to senior citizens to purchase a particular Medicare supplement, life insurance policy, denture adhesive, wheelchair, or any number of other products. Some viewers genuinely have a difficult time separating the characters who the actors play from the actors themselves. Obviously one who plays the part of a veteran attorney who is wise and brilliant has no particular credibility in the *real world* promoting a specific investment firm or certain brands of medical devices.

Einstein's credibility in the field of physics is enormous; however, he has virtually no authority in the field of psychology. Those who hold the title of *doctor* normally have expertise in their area, but care must be taken so that the *halo effect* is minimized. This effect occurs when people simply believe that a person's expertise or *halo* follows him wherever he goes. A doctor, a medical doctor that is, does not have any particular expertise in education; one who has a doctorate in education does not have any particular expertise in medicine. This does not imply that no transfer ever occurs. For example, a person with a Ph. D. in chemistry typically has significant expertise in math as well. But, generally speaking, expertise in one content area does not imply expertise in another content area.

In the earlier example, Tom Cruise, a superstar in the world of Hollywood, makes claims about a highly scientific and technical area of medicine. His claims about understanding psychiatry need to be balanced against the claims and research of thousands of medical doctors who have a keen understanding of psychiatry, medical research, and science in general. The vast majority of genuine experts in the field suggest that Cruise's understanding is superficial, and in many cases, is outright wrong.[13, 14]

His understanding and expertise do not compare to that of those who have medical degrees in the specialty and who have devoted their entire lives to working with researchers and patients. Accepting the medical claims of someone whose expertise is in acting, and not in medicine, should be a cause for concern. The point of this fallacy lies in believing what a person says, not because of expertise, but because of one's popularity or celebrity status.

Unfortunately, the masses of people often do just that. Not surprisingly, evidence indicates that we are more persuaded by claims when articulated by attractive, likeable, or popular individuals. We need to be especially careful about rejecting claims that come from unattractive people, but accepting of the same claims when they come from attractive people. Of course, people often demonstrate this bias without being aware of it.

The vast majority of people do not have the expertise to really evaluate the legitimacy of the scientific basis of evolution, for example. As a result, most people correctly accept the validity of evolution because it is the informed opinion of respected and educated experts in the field of biology.

It should be noted that the argument, *scientists once thought that the sun revolved the earth,* as a basis to accept the claims of nonexperts over experts is a weak argument—and actually it is sometimes even identified as a fallacy. Scientific or medical experts can be wrong, but the nature of the scientific method of evidence and reasoning usually finds mistakes sooner or later. Generally speaking, genuine experts are much closer to knowing the truth about the topics in their specific areas than are nonexperts. Nearly everyone can think of hundreds of situations when the experts were initially wrong, but using these exceptions to form a generalization that experts are wrong more often than nonexperts across the spectrum is actually a form of faulty thinking. Although it is also true that there are some social, religious, and political topics on which experts cannot agree, this lack of consensus is not grounds to suppose nonexperts have more authority.

Actors, athletes, celebrities, promoters, and other *so-called authorities* often try to promote some product or belief despite the fact that they have no expertise, knowledge, or experience with what they are endorsing. A popular and highly religious person who preaches that scientific evolution is a myth should not be viewed as credible if he has no comprehensive understanding of science or biology. Extreme care must be taken to shield oneself from this fallacy, simply because many famous people often speak about issues in which they have no credibility. Unfortunately, some people do, in fact, lend credibility to those who have been successful in getting their names in the newspaper or their faces on TV.

In the last decade many TV *infomercials* dealing with natural cures for physical ailments have become enormously popular. The vast majority of these shows are scandalous propaganda. Sometimes the promoter may not necessarily claim that he himself has medical expertise, but he does suggest that he has the authority to determine which doctors and what research to believe. In most cases, real experts with genuine credentials, education, and experience often despise these promoters or celebrities.

Should we believe such a lone promoter over the consensus that emerges from tens of thousands of doctors and medical researchers? Millions of people do exactly that. To substantiate the claim that there is a highly effective natural cure for virtually every physical ailment and medical condition, a promoter cherry-picks medical or nutritional studies to confirm a preconceived idea. That is, he reports on the scant evidence supporting his claim and ignores the massive evidence discrediting his claim.

Promoters of such claims or books cite numerous scientific studies *proving* their claims. They dig for evidence, regardless of how weak it is to support their claims and reject opposing evidence, regardless of how strong it is. Often they also unconsciously seek out only those doctors who confirm their claims. Such promoters frequently have little or no medical or research expertise and are not even qualified to determine who are credible doctors and what research is valid.

Often the *proof* comes from small-scale or inconsequential studies. A common joke among researchers asserts, *after treatments with the drug, one-third of the mice were cured, one-third died, and* ***the third mouse escaped.***

Testimonials from satisfied customers, one of the most biased and weakest forms of support, are used extensively. With millions of individuals, it is not difficult to find hundreds of enthusiastic individuals who say the so-called cures *really work.* Of course the millions of individuals who found the *cures* ineffective or even damaging are typically the vast silent majority.

The criticism above does not suggest that there are no natural cures or that conventional doctors despise all natural cures. In fact, the majority of doctors fully endorse the most fundamental of natural cures—*eat nutritional foods and exercise.* Obviously there are other natural remedies that have some effectiveness, but the wholesale claim mentioned above that *there are highly effective natural cures for almost all medical conditions and physical ailments* appeals to our emotions, not to our intellect. Substituting emotional and wishful thinking for logical reasoning can lead us down an expensive road that ignores science and critical thinking. The consequences can be deadly.

Earlier in this book, there was a brief discussion about the difference between being skeptical and being cynical. We must be careful not to become so skeptical that we become cynical. When we do that, we begin to reject that which should not be rejected and begin to accept that which should not be accepted. Being skeptical of conventional medical findings and their recommendations—that is, asking serious questions about validity, reliability, testing, and research methodology is encouraged. But being skeptical does not mean rejection. Just because we are somewhat dissatisfied—or find errors or deficiencies in mainstream medical and scientific research—does not mean that we should quickly endorse unusual or fringe medical treatments or approaches. In our attempt to become skeptical, we must be careful not to reject practices that are only partially effective in favor of other practices that are even less effective—or even dangerous.

In review, the fallacy of *ad vericundiam* appeals to and exploits our psychological weaknesses. In fact, this appeal is so powerful and common that only a conscientious application of critical thinking can resist its influence. Simply put, we want to believe what certain people say despite their lack of authority. When the evidence is too difficult to understand, or when the claims are too technical to comprehend, consulting an authority—an authentic authority in the field— is the next best option for determining what to believe.

Chapter 6

Fallacies of Language

Emotive Words

Equivocation

Accent

Personification

Composition

Division

Amphiboly

Chapter 6

Fallacies of Language

> *The difference between the right word and almost the right word is the difference between lightning and the lightning bug.*[1]
>
> *—Mark Twain—*

In many ways, this quote captures the essence of this entire chapter on language fallacies. Language is incredibly powerful, and it can be manipulated to fool people into accepting claims that are false, but that appear to be true.

The popular quote, *the pen is mightier than the sword,* reminds us that carefully chosen words can be very motivating. Words that persuade us are often more powerful than swords that threaten us. That is, words are persuasive tools that change genuine internal beliefs. In contrast, physical threats may force us to temporarily change our behavior, but rarely do they lead to authentic long-term changes in belief.

What we call something, or its label, is like the wrapping paper on a present; it is not always a good indication of the true contents of the box. The enticing words used by persuaders in making a claim are equally deceptive. The wrapping paper, or the fancy bow on the top, grabs our attention and entices us to accept a present, just as words can seduce us to accept a claim.

As suggested in Chapter One, words are like filters, or lenses, that influence and shape how people actually record reality. Changing lenses on a camera filters out, or magnifies, certain aspects of the world so that the recorded realities vary depending upon the lens used. Wearing a new pair of glasses can change the focus of our world just as words can alter what we see and believe.

Words are sometimes intentionally manipulated to gain an unfair advantage. But sometimes people have simplistic notions about certain labels. For example, most people hear the word *convict,* and then conjure up an image of a violent person. Most *convicts,* however, become *convicts* not because of violence, but instead because they were caught stealing or buying illegal drugs. When labels are used, it is important to look beyond the one-dimensional portrait that this can paint and instead ask whether the label provides a comprehensive understanding.

For example, many people are *conservative* on economic or fiscal issues, yet are *liberal* when it relates to social or cultural issues. Labeling such individuals as *conservatives* or *liberals* is inaccurate. Labels simply do not allow the complexity and individuality of people to be easily seen. As mentioned previously, this book is being written during the 2008 presidential campaign season. Presidential candidate, Senator John McCain, for example, tends to hold some views that are considered somewhat liberal, although he holds many other views that are conservative. This *hybrid* candidate troubles some in his party who are strong conservatives. On the other hand, this mixture of views may be more attractive to moderate voters in the general election. President George W. Bush, a strong *social conservative* on most issues, often does not apply that same philosophy to fiscal matters. Whereas some people will label another as a *liberal* or *conservative* just to incite some positive or negative emotional response, others employ the terms in ways that shows their simplistic or stereotypical understanding of political ideologies. Again, using broad-based labels to describe these individuals or positions attempts to take complicated opinions and oversimplify them.

Other times fallacies are due to the inherent ambiguity or vagueness in our language. Before going shopping, my wife used to say, *I'll be back in a little while.* I assumed that she meant an hour or two. I quickly learned that her definition of *a little while* usually meant four to five hours. After a first date, the man might say: *I'll give you a call,* meaning he might call within a week or two. *Within two days* is often his date's definition. Such ambiguity of the language can obviously create problems.

Another way people misuse language occurs when they present their opinions as if they are factual statements. Nearly everyone will recognize that the statement, *dogs make better pets than cats,* is purely a subjective opinion. On the other hand, sometimes making the distinction between a fact and an opinion is more difficult. For example, I recently heard someone say, *because taxes are too high, we should vote for Proposition 105.* Part of this statement, *we should vote for Proposition 105,* is an opinion. However, even the initial premise of the argument, that *taxes are too high,* is not a factual statement, but a subjective opinion.

Similarly, people often make statements as if they have already been proven as facts. For example, stating that *the low morals in this country are because people no longer believe in God,* provides two statements that are assumed to be true. Is it factually correct to state that there are low morals in this country? Is it factually correct to say that people no longer believe in God? When people provide their opinions as factual statements, we should be ready to challenge their claims.

These are just a few of the many ways that language is used inappropriately. Obviously not all such errors have been given specific fallacy names. This chapter, though, identifies a number of errors that have been specifically labeled as common language fallacies. In some cases, persuaders use words unfairly to rile up emotions and suppress thinking. In other cases, the vagueness of words, sentences, punctuation, or grammar allows persuaders to gain an advantage in making a claim.

Fallacy

of

Emotive Words

Pollster asking people on the street who they are going to vote for:

Are you going to vote for a candidate who will take away even more of your hard-earned money with massive tax increases and then waste it on programs that don't work?

Are you going to vote for a candidate who is an arrogant, out-of-touch liberal?

Are you going to vote for a candidate who will coddle terrorists and bow down to their demands?

Are you going to vote for a candidate who has been a life-long Washington bureaucrat who wants to send another 3,000 kids to their deaths in Iraq?

The *fallacy of emotive words* is a faulty way of thinking, in which one seeks an unfair advantage by using emotionally charged words to bias opinions.

In looking at these questions, 95% of the respondents will answer, *NO!* The remaining 5% will probably refuse to answer the question. Most critical thinkers would be in this 5%. That is, they would refuse to respond because the questions are so biased and emotionally-charged that the responses are entirely invalid.

There are legitimate ways of using language to justify claims and therefore cause changes in opinions. But, there are also many manipulative and sneaky ways of massaging words to gain an unfair advantage. Simply labeling an idea with an emotionally charged word can make people accept or reject it without considering the merits of the idea.

The words themselves can influence the realities that others internally record in their minds. Stating that the United States *invaded* Iraq creates a different view of reality from saying Iraq was *liberated* by the United States. An *invasion* implies an illegal and unwarranted use of military forces by a hostile country towards another. On the other hand, *liberation* implies freeing citizens from a repressive government. Those in favor of the war call it *liberation,* whereas those opposed to it call it an *invasion*. Saying 3,000 lives were *sacrificed* evokes a different reaction than saying 3,000 lives were *wasted*. One word can create a positive or negative emotional reaction. Obtaining support for a particular issue depends upon the ability to select words or phrases that create the desired realities in the minds of the receivers.

Politicians might be successful in arguing that *estate taxes* should be increased; however, they might be less successful in arguing that *inheritance taxes* or *death taxes* should be increased. In truth, these three terms actually describe essentially the *same* tax. Simply using a different term to describe a concept actually influences how people perceive the reality of the concept. In fact, getting a bill through Congress and through the court of public opinion depends, to a large extent, on the label the bill is given. The contents of two bills can be identical, but the public might oppose one bill and support the other. Whether it is the title of a bill or a headline in the newspaper, a few emotionally laden words can often defeat scores of logical arguments and piles of evidence.

As will be explained in another chapter, persuaders are often effective in getting people to replace logic and evidence with feelings. Many terms, called *emotive words,* are those that conjure up some emotional reaction from the listeners. Such words can be harmful because they often cause people to make decisions with their hearts and not with their heads. This is not to suggest that emotions are never relevant to decision-making. In fact, using one's heart to make some decisions is entirely appropriate. The fallacy occurs when people believe they are making a primarily intellectual decision, when, in fact, manipulative words trigger their emotions to make that decision.

Persuaders know instinctively that many people can be more influenced when their emotions are aroused. We must be careful of such words because they have a built-in emotional bias. Are you in favor of this *bloated wasteful liberal tax?* This survey question is peppered with emotive words that entice people to respond negatively. Are you in favor of electing *a trigger-happy, ultra-conservative, right-wing Republican?* This too generates negative emotions.

Politicians from all political parties try to bias the voters by using such emotionally charged words. Effective persuaders know the value of emotive words; however, critical listeners should be skeptical and realize they are the targets of manipulation. Unfortunately many people rarely even realize that certain words tickle their emotions in ways that lead them to vote in a certain way.

While some fallacies fall squarely into specific chapters in this book, this one actually overlaps with others. It is certainly a language fallacy, but it could be placed in the chapter on *relevance* as well. The *relevance* chapter looks at the various ways that people use irrelevant elements when making a decision. Sometimes emotions are irrelevant elements.

Fallacy

of

Equivocation

> *During class today I will be asking you to donate to **Charity**. Please contribute what you can.*

The *fallacy of equivocation* is a faulty way of thinking, in which one uses an alternative meaning of a word to deceive or win an argument. Such a fallacy operates by taking advantage of multiple meanings of some words.

In the example above, most of the students would assume that *Charity* refers to an organization that provides aid to those in need. But *Charity* can also be a woman's name. Whereas students believe they are giving to those in need, the money I am collecting is actually going to this woman, *Charity*. Letting people believe one definition of a term while secretly using another legitimate definition allows one to gain some advantage without being entirely untruthful.

An advertisement in a men's magazine, for example, claims they will provide a *dozen highly selective dates* for a mere $300. After paying for this service, a week later a box of dates, the edible kind, is delivered to the man's front door.

Likewise, in England an obese woman reads about a weight-loss program in a tabloid magazine. The claim is that anyone who buys this program is guaranteed to lose 30 pounds in a month. Enticed by this guarantee, she purchases this program with her credit card. The program, she is told, will arrive shortly. A few days later she receives a letter with a receipt indicating her account was charged *thirty pounds.* A certificate is also included that reads, *Congratulations, you have just lost 30 pounds.*

Similarly, a businessman frequently gets a call on his cell phone in the evening from his wife. She wants to know where he is. He replies, *I am still at **the office**, and I'll probably be here a while.* The wife, unfortunately, does not know that a popular sports bar is called *The Office.*

Equivocation can also occur in the doctor's office. An adult daughter takes her mother to the doctor for a physical exam. The daughter grills the doctor after the exam about every possible aspect of her mother's condition. Because there is a history of early heart attacks in the family, the daughter asks about this. The doctor replies, *your mother has a good heart.* The next day the mother dies of a massive heart attack. The doctor claims he meant that she was a *good-hearted,* kind woman.

This fallacy is also sometimes obvious in court cases. In many cities, the highway includes a lane known as the High Occupancy Vehicle (HOV) lane. This less-congested, faster-moving HOV lane is reserved for those who are car-pooling, or who have at least two occupants in the car. There was a case of one driver, a pregnant woman, who was stopped and given a ticket because she was the only occupant in the car. She attempted to use the fallacy of equivocation by arguing that, because she was pregnant, there were actually two occupants in the car. Her case went to court. She lost.[2]

As the foreman of a jury in 1994, I was somewhat uncomfortable with the task given to us—if the *preponderance of evidence* shows fault, find the defendant guilty. Some jurors thought this meant *most,* while others suggested it meant the *vast majority,* while others said it was somewhat less than *beyond a reasonable doubt.* Because we could not agree, we asked the judge for a definition. His response was that *it was the jury's job to define what it meant.*

In some cases, speakers knowingly use the ambiguity of language for deception. In other cases, however, there is not a deceptive intent. For example, millions of Americans read medical studies about the effectiveness of diets, medicines, and treatments. They diligently analyze the statistics to determine which drug, for example, is better than the others. They select a drug, for example, because research has found it to be *significantly* more effective than another drug. Researchers use the word *significantly* in a highly technical and statistical way.

In contrast, the public simply assumes any drug that has been shown to *significantly* improve symptoms or help a *significant* number of people is likely to dramatically help the majority of individuals. This is entirely false. In most research, *any* effect that occurs *beyond random chance* is typically considered *significant.* For research purposes, this is often set at five percent. In other words, if a drug reduces symptoms for just five of 100 people, it is advertised as being *significantly effective.* And even in these cases, just a five percent improvement in each of these individuals is known as *statistically significant.* So, a drug that has been shown to be *significantly effective* may have no effect what-so-ever on the symptoms for 95% of the people, and for the remaining 5%, their improvement could be as small as 5%. In some cases, the improvement could be even less.

This is not to suggest that the differences are always that small, but frequently the differences are far less than most people think. The words *significant, beyond a reasonable doubt,* or *preponderance of evidence,* mean dramatically different things to different people.

The previous scenarios are rather obvious examples as to how two meanings of a word can create confusion. However, most examples of these fallacies are far less obvious and far more dangerous.

In parts of the United States, a few school boards are caving in to pressure from some passionate citizens who seek to replace the teaching of *evolution* with *creationism* in science courses. *Creationism,* as defined by those on the extreme end of Christian fundamentalism, is entrenched in a belief that every word in the Bible is literally true. The firm belief is that God, for example, created the world in six days. To be fair, most fundamentalists actually don't seek to completely replace evolution, but rather to present creationism as another equally valid scientific theory. But, this Christian religious story from the Bible, just like religious stories from other holy books, is not science. The Koran, the holy book of Islam, and the Veda, the scripture of Hinduism, provide different stories about the origin of life. Science has very specific and exacting principles of observation, measurement, reliability, and validity. Many parts of evolution have been proven with standards that approach virtual certainty. Religious stories do not have these qualities.

Suggesting that *evolution* and different *creationism* stories from various religions are all *scientific* demonstrates a massive lack of understanding about basic principles of biology and genetics. The stories about the beginning of life, regardless of what religion they arise from, should be respected. They are part of culture and history. It is certainly appropriate to examine these beliefs in a religion, philosophy, or humanities class, but not in a science course.

Many courses don't seek to obtain the truth by using logic, reasoning, and evidence—as science does. Some courses simply examine people's beliefs without asking for proof. Discussing different stories of creation or debating the meaning of life is beyond the realm of science. Suggesting that science textbooks include both evolution and creationism as valid scientific principles is a false analogy. One is based on science, and the other is not.

Whereas the dominant religions of the world are usually concentrated in certain parts of the world, scientific principles such as evolution are not limited to any race, sex, culture, or religion. Scientific principles are used to discover natural explanations for observations. On the other hand, religion does not focus on natural explanations.

As a way of forcing the teaching of creationism, in 2005 the Kansas State Board of Education, in effect, changed the definition of *science* by removing the words, *natural explanations* of the world.[3] This change would now allow teachers to use *supernatural explanations* of observations and events. Although many people don't realize the importance of this change, such a move actually eliminates *science* from the *science* curriculum. That would be much like having medical schools teach that voodoo, hexes, curses, magical spells, and witchcraft are legitimate tools of modern scientific medicine.

By allowing *supernatural explanations,* it could be argued that virtually any explanation could now be included in *science.* Rather than explain that shifting plates on the earth's surface caused the earthquake, a teacher could conjure up any explanation—perhaps a god, a devil, a witch, or even an angry herd of invisible dinosaurs caused the earth to crack.

Fundamentalists believe their beliefs are true, and they want others to be taught their *truth.* But truth is not simply determined by whether individuals believe something or not. There are many other ways of determining truth by using scientific principles and critical thinking. As discussed elsewhere, in many cases there is virtually no relationship between belief and truth. Centuries ago, virtually everyone believed with certainty that the sun revolved around the earth.

Science has an excellent reputation for ultimately getting to the truth, and including the belief of *creationism* in a science curriculum would be a first step toward promoting creationism as a universal truth. However, creationism is not science, and most scientists refuse to label it as such. In an attempt to persuade school boards that students should be taught this belief as a truth, the school board members simply changed the definition of *science.* This is perhaps one of the most blatant attempts in decades to achieve a goal by employing the fallacy of equivocation.

When people around the world are asked whether they believe in *God,* most say they do. But, when this question is asked in India, their *God* is not the same *God* of most Americans or Egyptians. There are three dramatically different meanings of *God* just in these three countries. Because of the ambiguity of this word, the language makes it appear that the people in these countries believe in the same thing, when, in fact, they believe in radically different notions.

Altering the meaning of a word allows an unfair advantage in many cases. Because the United States agreed to the terms of the Geneva Convention, it agreed not to ***torture*** prisoners of war. The war in Iraq that began in 2002 obviously produced thousands of prisoners of war. Some of these prisoners were no doubt terrorists, and obtaining key information from them was a high priority. *Inflicting intense pain* on prisoners or *making prisoners believe their deaths were imminent* is the standard definition of torture.[4,5] If this definition is used, evidence indicates that torture was clearly used in some cases. However, in 2005, the president repeatedly said that *the government does not torture people.*[6]

Of particular focus was the use of a technique known as *waterboarding* to obtain information from prisoners. This technique makes the prisoner believe he is dying. It creates the physical and mental sensation of drowning. For those interested in deciding for themselves how to label this technique, there are a number of Internet videos that are very explicit. This involves blindfolding a prisoner, strapping him on his back to a bench, and stuffing his mouth full of rags. Then buckets of water are poured on his face and head so that the rags become full of water. Water then trickles down his throat and causes a gag reflex. Although this typically does not cause death, those undergoing this treatment cannot escape the physical sensation and mental belief that they are drowning. White House authorities asserted that this technique was legal and did *not* constitute *torture.*[7]

The point is not whether torture is effective or not—many argue that it is an essential method in fighting terrorists. The fallacy, however, occurs because one is gaining an unfair advantage by refusing to accept the standard definition of a term, or by creating an entirely new definition.

Rather than calling *waterboarding* torture, the procedure was renamed—*enhanced interrogation techniques.*[8] Prisoners of war were also given a new name—*illegal enemy combatants.*[9] While there are laws that govern the treatment of *prisoners of war,* there are no specific laws dealing with *illegal enemy combatants.* As a result, the claim was that no specific law prohibited using the types of techniques that were employed against those who were imprisoned.

Obviously, all political parties use this type of deception on a regular basis. Presidential candidate Hillary Clinton in 2008 used the fallacy of equivocation by saying she did ***not*** technically vote for the war in Iraq, a war that has become very unpopular with most Americans. She said she voted for an *authorization for the President to use military force* against Iraq in 2002. That is, she said she voted to give *authority* for the President to use force, but that she did not think he would or should actually use military force and start a war.[10] Many people, however, believe she knew that *authorization* was clearly a *vote for war.* Obviously it is impossible to know with certainty how she was personally defining these terms, but it appears a new definition, a more acceptable definition, was created.

The *fallacy of equivocation* was used in all these cases to gain an unfair advantage. Candidates from all political parties use this tool of persuasion to accomplish their goals. Whether used to gain a political advantage or to avoid charges of illegality, fallacies of equivocation are frequently employed.

Fallacy

of

Accent

Wife:	*Does the dress make me look fat?*
Husband:	*No, the* ***dress*** *doesn't make you look fat.*

The *fallacy of accent* is a faulty way of thinking, in which one alters the intended meaning of a sentence or question by using an inappropriate vocal inflection or accent to make a claim.

A previous fallacy, *equivocation,* alters meaning by using a questionable definition of a specific word. The *fallacy of accent* also creates confusion, but it does so by vocally emphasizing a specific word in a sentence to alter its intended meaning.

In the example above, the husband is using this fallacy to express his displeasure with his wife by using an accent that turns a compliment into an insult. If there is an equal emphasis on each word, the statement is actually a compliment. But, if the husband emphasizes the word *dress,* he is insulting his wife. This accent implies she *looks* fat because she *is* fat. If the wife becomes upset at the husband for his apparent insult, he can deny that he meant it as an insult. A different accent creates a different meaning, and he can simply use the definition that is most beneficial to him.

Suppose a wife accuses her husband of *sleeping with another woman.* A cheating husband might deny this charge by exclaiming: *I am not* ***sleeping*** *with another woman.* Technically he is not *sleeping* with her—he has sex with her and then goes home.

Accent can be used for advantage in the courtroom. Suppose an attorney is questioning whether a defendant was ever convicted of a previous crime. The defendant claims, *I was never convicted of a crime.* Later, however, the attorney might damage the defendant's credibility by repeating that same statement, but with a slight shift in accent—*the defendant claims he was never* ***convicted*** *of a crime.* In effect, the attorney just suggested that the defendant was probably *charged* with a crime, but was technically never convicted of a crime. Such a maneuver can create doubt about the defendant's past. Even the remark, *he never* ***confessed*** *to the crime,* suggests he is actually guilty, but simply refuses to admit to his guilt.

On a more personal level, my daughter often uses the word *dude* in her conversations. However, changing inflections in her voice produces three different meanings. In one case, she says the word quickly, but with an even tone, and it basically means *hi* or *hello.* In another case, she draws out the word with an upward inflection at the end, and it means *pay attention.* Finally, she draws out the word even more and it essentially means *I agree.* One has only to recall the inflectional style of a *valley girl* or *surfer dude* to realize how accent can change meaning.

Sometimes an accent is used legitimately to communicate a particular meaning. The fallacy occurs, however, when one uses vocal accent to create an alternative meaning in order to gain an unfair advantage in making a claim.

Fallacy

of

Personification

> *All living things have emotions. I know my flowers and plants become sad and don't grow well when I don't talk to them. Even the goldfish in my fishbowl swim slower and become depressed when there is a tragic story on the TV news. We have a law against emotional abuse of other humans. We need to have such a law to protect our plants and animals.*

The *fallacy of personification* is a faulty way of thinking, in which one attempts to make a claim by assigning human emotions or traits to plants, animals, products, or concepts.

The foregoing example is obviously an exaggeration, but it does expose the rather odd reasoning of a small segment of our population. This example is not intended to dismiss those who lobby for the humane treatment of animals. But, suggesting that most animals or plants have emotions in the same way that humans have emotions is a claim without much support.

Suggesting that ***some*** animals might experience a few basic emotions that are somewhat similar to raw emotional reactions in humans may have some limited merit. For example, we assume that injured animals experience some sensations that are similar to the sensations people experience when they are physically in pain. Humans sometimes moan when they have pain, and perhaps moaning sounds that come from animals are an expression of a similar sensation. Again, this may have substantial merit, but assuming that animals, especially the lower forms, have the wide range of complex emotions of humans is not justified.

We all are familiar with the human emotions of love and lust, and we can identify certain behaviors that provide evidence for such emotions. Humans may express an emotion of love or lust by having sex. Do bumblebees that mate have similar emotions? Such an assumption is dubious at best.

A great deal of animal behavior is probably devoid of any feelings and emotions. In looking at the behavior of ants or bees, for example, prewired genetic traits cause these insects to simply respond to certain events. The behaviors typically do not occur from any rational or deliberative process, but rather from many programmed instinctual responses.

This fallacy is often known as *the Walt Disney fallacy.* In Disney productions, many animals, and even some plants, have been animated in ways that seem to give them genuine human emotions and behaviors. Mickey Mouse and Donald Duck are perhaps the most popular examples of personification. Intellectually we know that mice and ducks do not possess the scores of emotions expressed by the cartoon characters. But, at an emotional level, we have to wonder whether we transfer some of Hollywood's personification magic to our pets and animals.

Certainly writers will sometimes use personification for dramatic effect; however, they are not suggesting that plants, for example, really have emotions. *The sunflower was aching for water and turned her new baby blossom to the shade to protect her from the sting of the vicious solar rays.* The human qualities given to this plant are simply used to help readers understand the perceptions of the writer.

Obviously those people who refer to *Father Time* and *Mother Earth* are not committing this fallacy because they know those are simply expressions. Similarly, giving a woman's name to a hurricane or ship does not literally suggest human qualities.

Language can be used to incorrectly suggest that most animals and plants have emotions, opinions, or attitudes that are extremely human-like. Given this so-called similarity, it is then implied that such plants and animals should be treated in the same way that humans should be treated. Although this fallacy is less common than many fallacies, it occurs every time we allow someone to project human qualities onto an animal or plant for the purpose of winning an argument.

Fallacy of Composition

I know all the individuals in the group are very intelligent. Therefore I am sure that the group itself will make very intelligent decisions.

The *fallacy of composition* is a faulty way of thinking, in which one automatically believes that the characteristics of the parts that make up the whole have the same characteristics as the whole itself.

Perhaps the best way of remembering this fallacy of composition is to recall the process of writing an essay or composition. A writer will take parts (paragraphs) and put them together to compose an essay.

When one is writing an essay, each paragraph might be very coherent; however, simply pasting these paragraphs together does not mean the essay itself will be coherent. If coherent paragraphs are not connected appropriately, the essay itself will be incoherent and disjointed.

Similarly, putting highly intelligent individuals together in a group does not mean they will be cohesive enough to produce intelligent results. In fact, social psychologists have found that this phenomenon occurs so frequently in groups that it has been the focus of much study. Sometimes internal personal conflicts will actually cause the most intelligent and informed individuals to compete against each other. At times more outspoken members will intimidate others so that dissenting opinions are never discussed. High individual IQ does not necessarily mean high group IQ.

The USA men's Olympic basketball team arguably was composed of the best basketball players in the world in 2004. As a team, however, they earned only a bronze medal, a major disappointment. As the fallacy of composition suggests, the quality or characteristics of the individual parts (or players) does not necessarily equate with the quality of the whole (or the team). In 2008, however, the best players showed they were the best team by winning the gold medal for the USA.

On a more concrete level, hydrogen is a gas and oxygen is a gas. Putting these two gaseous molecules together does not create a gaseous product. Instead, water is created.

In education, each professor is a specialist in a particular field. But, does that mean a group of professors is simply a large collection of specialists who are unable to solve a general problem? Quite the opposite is true. Often specialists collaborate to solve more general problems.

The same applies to medicine. A patient who suffers from a number of general physical complaints and symptoms may find little relief from simply seeing a single physician. However, a group of specialized doctors, when working on a case together, can often combine their expertise in a way that produces quality results. That is, the group is greater than the sum of its parts.

One interesting example of this fallacy focuses on a common, but incorrect, assumption that most stocks increase in value over the long haul. In fact, *most* stocks actually *lose* value—even over the long run. Over the last 30 years, 75% of the stocks have gone *down* in value (about 2%). However, the remaining 25% of the stocks rose so much that, in general, the stock market increased substantially in value.[11] So, generally speaking, it was wise to be invested in the stock market. That is, the losing characteristic of most stocks did not mean the stock market as a whole also lost value. Of course, one does not know in advance which stocks are likely to go up and which stocks are likely to go down. Being well diversified, or owning a wide variety of stocks, increases the chances of also owning those stocks that do especially well.

This fallacy of composition does *not* suggest that the characteristics of the parts *never* have same characteristics of the whole. For example, the individual blades of grass might be green, and the lawn that is created by the blades might also be green.

On the surface, it may not be apparent why this fallacy is considered a language fallacy. Historically, it has been put in this category because the language itself creates ambiguity, and that ambiguity often leads to mistaken conclusions. For example, if I say the *deer caused the accident,* it is not clear whether I was referring to one animal or a group of animals. *Deer* is one of those few words to describe a single animal as well as a collection of animals.

In our situation, the ambiguity frequently results because the writer doesn't make a distinction between words such as *every, all, some, each, and entire.* For example, just because *every* person in a group of five individuals is unable to get over a nine-foot wall does not mean *every* group of five individuals cannot get over this wall. Put another way, although *each* person is unable to get over the wall, *some* groups can, in fact, get over the wall. Being hoisted up on another person's shoulders would allow most individuals to get over. So, even when every single person has a specific characteristic, this does not mean the *entire* collection of these individuals will have the same characteristic.

The next fallacy, *division,* is the converse of *composition.* Whereas the *composition fallacy* reminds us that the qualities of the parts may not show up in the qualities of the whole, the *division fallacy* proves that what may be true about the whole may not be true about the parts that make up the whole.

Fallacy

of

Division

> *I see that the Federal Reserve lowered the prime interest rate again. This is certainly good news for Americans.*

The *fallacy of division* is a faulty way of thinking, in which one automatically believes that the characteristics of the whole have the same characteristics of the parts that make up the whole.

Put another way, what is true of the whole is not necessarily true of the parts that make up the whole. A brick wall, as a whole, is visible to the naked eye. On the other hand, not a single one of the trillions of atoms that make up the brick wall is visible to the naked eye. The characteristic of the parts, invisibility, is not a characteristic of the whole.

For example, it is possible that a team, as a whole, is extraordinary. But, when we look at the individuals on the team, not a single person is extraordinary. Similarly, water does not burn, but hydrogen and oxygen, the parts that make up water, both burn.

On the whole, Americans will probably welcome lower interest rates. But, what is true for Americans in general is not necessarily true of the various segments that make up the population. People who are buying homes, investing in the stock market, or purchasing new vehicles would probably enjoy lower monthly payments or better returns when interest rates are low. On the other hand, retired senior citizens frequently find lower interest rates undesirable. Because the retired need dependable sources of income, they often rely on extremely safe investments, such as savings accounts or bank certificates of deposit. These types of investments pay a guaranteed rate of return, but during times of low interest, their accounts also earn very little.

In general, homeowners in many major metropolitan areas in the United States saw their home values appreciate substantially from 2000 to 2006. Again, on the whole, many people saw this as good news, but for an individual who was attempting to buy a house during that time, the rapid appreciation obviously translated into higher monthly mortgage payments.

In 2006, many of these same housing markets saw a sudden and drastic reversal. Housing prices plummeted, and homes that had normally sold in 30 days were on the market for many months. (Note: See *fallacy of regression* for a possible reason for this sudden shift). This was not good for the economy as a whole, but for those who had excellent credit or available cash, it was a buyer's market.

Gas prices in the United States increased rapidly during the first half of 2008, and this was damaging to the economy and to the population as a whole. However, for some who invested heavily in oil companies, the surging gas prices resulted in very lucrative investment returns. Higher gas prices also increased sales of hybrid vehicles as well. So again, what is true for the population as a whole is not necessarily true for the individuals who make up the population.

Sometimes there is sufficient evidence to conclude that what is true for the whole is also true of its parts. For example, the brain will die if it not supplied with oxygen, just as the neurons that make up the brain will die without oxygen. A fallacy is committed only when there is an automatic and insufficient assumption that the whole and its parts have the same characteristics.

Applying national statistical data about a whole to an individual person or family shows clearly how this fallacy can produce ridiculous conclusions. For example, suppose a census indicates that the population of a particular country is 51% female and 49% male. Obviously, this does not mean that the typical person on the street is *just a bit more female than male.* Also, just because there are 2.5 automobiles per family does not mean that the typical family has two complete cars and another half of one lying around somewhere. Likewise, on the whole, suppose that the average high temperature in Phoenix for the twelve calendar months is 75 degrees. Clearly it would be questionable advertising to suggest that Phoenix enjoys a *balmy climate of 75 degrees.*

Similarly, one might wonder how anyone could drown in a river that averages only a foot deep. Such a river, of course, could have many places where the water is more than six feet deep. That is, the river's shallowness as a whole does not mean all parts of the river are also shallow.

Like the previous fallacy, it may not be apparent why this is considered a fallacy of language. Because one incorrectly assumes that the whole is the same as the parts, this fallacy could arguably be put in a category of faulty presumption—another chapter in the book. But, the key point of confusion is created in how one uses language. The word *fish,* for example, is used as a verb as well as a noun. And when used as a noun, the term itself doesn't identify whether one is referring to a single fish or to several fish. In the same way, the writer or speaker frequently uses words that make it unclear whether he or she is referring to all cases, some cases, each case, or every case. A claim, *aspirin produces bleeding,* is ambiguous. Is the claim that *all* aspirin produces these effects, or that only *some* specific level of aspirin use creates these effects? Is this side effect evident in *every* case, or just in *some* cases? Sometimes this ambiguity is used intentionally to gain some advantage. But frequently the inherent and imprecise nature of language creates this confusion.

Fallacy

of

Amphiboly

Man to a woman at a bar: *If I told you that you have a beautiful body, would you hold it against me?*

The *fallacy of amphiboly* is a faulty way of thinking, in which one uses a poorly worded or structurally ambiguous sentence to make a claim or win an argument.

This fallacy is very similar to equivocation. Equivocation, though, centers on various meanings of a single word. In contrast, an amphiboly centers on various meanings of the entire sentence due to its wording, structure, grammar, or punctuation.

In the example just given, the wording suggests two potential meanings. Is this question asking whether she would hold a grudge against him for such an insensitive or provocative statement? Or, is he asking whether she would be willing to hold her body against his? If she responds with disgust, he can simply say, *oh, that's not what I meant.* Amphibology allows one to use whatever meaning is most convenient.

How a sentence is worded and interpreted can clearly have major political consequences. 2008 Republican presidential nominee, John McCain, said that it was clear that a particular terrorist organization in the Middle East wanted the Democratic presidential nominee, Barack Obama, to win the election.[12] In response to this insult, Obama stated, *And so for him to toss out comments like that I think is an example of him* ***losing his bearings*** *as he pursues this nomination.*[13]

Most people interpreted Obama's comments to suggest that McCain's campaign was getting off-track or losing its direction. McCain's staff, on the other hand, was insulted because they perceived the phrase, *losing his bearings,* as an attack on their candidate's age and mental condition.[14] The wording allowed enough ambiguity that two people could interpret the same message in drastically different ways.

Perhaps the definitive political example came from President Truman. When asked whether he was going to run for a second term, Truman was reported to have said, *I do not choose to run.*[15] Was he saying he was not going to run? Or, was he saying that he preferred not to run, but that his country or obligation was forcing him to run again?

Like Truman, many politicians today frequently insert ambiguity into their messages. Suppose a politician is asked whether she is in favor of a tax increase to balance the budget, and she responds: *I am not in a favor of a tax increase that would hurt small businesses.* What exactly does this response mean? Does it mean she is not in favor of a tax increase *because* it would hurt small businesses? Or is she saying that she would accept a tax increase to balance the budget, as long it doesn't hurt small businesses?

On a more personal level, some years ago I bought a CD player, and I had intended to give it to my wife for her birthday. However, I actually gave her some other gift and decided to return the CD player to the store. As I waited in the return line at the store, I listened closely to what the store clerk was telling the customers in line. To the first man, the clerk said that he could not give a refund without a receipt. Because the customer didn't have a receipt, he became disgruntled and walked out of the store.

The second person in line was told the same thing and precisely the same result occurred. I was ready to leave the store myself because didn't have a receipt either. So, when I got to the front of the line I admitted, *I got this CD player as a birthday present, and I don't have a receipt for it.* I assumed I would be told the same thing as the previous two people in line. Instead, I was surprised that the clerk said, *well, if you got it as a birthday present, you obviously don't have a receipt. Let me look up the price, and I will give you a refund.*

Before I understood what was going on, the clerk had put about $50 in cash in my hand. As I walked out the door, it occurred to me that I had just benefited from an amphiboly. Apparently when I told the clerk that *I got it as a birthday present,* he assumed I had *received* it as a birthday present.

Similarly, Moses Hadas said, *Thank you for sending a copy of your book—I'll waste no time in reading it.*[16] Does this mean he will begin reading it immediately? Or does it mean he thinks that reading your book is a waste of time, and therefore he is not going to read it?

Sometimes the confusion occurs, not from the wording of the sentence itself, but from how the sentence is punctuated. For example, consider the following hypothetical example from a will:

I leave John and Joe, and Jane the following: my house and my car.

What does this say? Does it mean that John and Joe will inherit the house and Jane receives the car? Or does it mean that all of them will receive both items? In this example, changing the punctuation alters the meaning dramatically. Obviously this is an extreme example; however, in other cases, the smallest structural or grammatical change can have grave consequences.

A poorly written will can lead to lengthy court cases and hard feelings between family members. Attorneys often use highly technical and legal language to compose important documents. Although such language may appear incomprehensible to most people, the wording, including its grammar and structure, is carefully composed to avoid confusion.

Many people assume that the words themselves transfer the intended meaning from the writer to the reader, and that sentence structure and grammar are of secondary importance. Although this is sometimes true, consider the often-told story about an English professor who wrote the following words on the blackboard and then asked the students to punctuate it correctly:[17]

A woman without her man is nothing.

The men wrote:

A woman, without her man, *is nothing.*

The women wrote:

A woman: without her, *man is nothing.*

There are hundreds of jokes about poorly worded accident reports.[18] Note how the wording makes the meaning very ambiguous.

The pedestrian had no idea which direction to go, so I ran over him.

I had been driving for 40 years when I fell asleep at the wheel and had an accident.

I was unable to stop in time, and my car crashed into the other vehicle. The driver and passengers then left immediately for a vacation with injuries.

I pulled away from the side of the road, glanced at my mother-in-law, and headed over the embankment.

The guy was all over the road. I had to swerve a number of times before I hit him.

An invisible car came out of nowhere, struck my vehicle, and vanished.

Likewise, a welfare department ***allegedly*** received the following statements from various individuals:[19]

I am forwarding a marriage certificate and six children. I have seven but one died, which was baptized on a half piece of paper.

I cannot get sick pay. I have six children. Can you tell me why?

I am forwarding my marriage certificate and five children, one of which is a mistake as you can see.

Amphibology can produce relatively harmless, even comical interpretations. In other cases the ambiguity can lead to disastrous results.

Chapter 7

Fallacies of Presumption

Slippery Slope

Appeal to Ignorance

False Charge of a Fallacy

Apriorism

Moving the Goalposts

Contradictory Premises

Either-Or

False Dilemma

Middle Ground

Circular Argument

Chapter 7

Fallacies of Presumption

Don't Assume! Most of us have been lectured on this principle, and most of us generally believe we are not guilty of making this error frequently. This is a faulty assumption. We unknowingly make unwarranted assumptions all the time. *Fallacies of Presumption* are committed when we make incorrect inferences or use questionable assumptions in our arguments.

Pulling apart our conclusions from the assumptions they are based on is as difficult as picking pieces of chewing gum out of a child's hair. It is difficult, tedious, and time-consuming. But, it is a necessary chore to ensure a reasonable result.

Unlike the gum in the hair, however, often we cannot even detect where the assumptions start, and the conclusions begin. For example, if I glance at the clock and then say, *it is 10:30,* most people would say I am simply reporting the facts—and not making any assumptions and inferences. However, I am actually making a number of assumptions. To be precise, I am observing that the little hand on the clock is between the ten and eleven and that the big hand is on the six. I am then assuming that it is 10:30 based on these observations. I am also assuming that the clock is working and is accurate. Although this process seems like it is splitting hairs, it is often necessary if one is to achieve a high level of critical thinking.

Suppose, for example, you receive a letter from your friend, Jack, who lives in Spain. Jack writes that he and Pat are getting married on 10-12-09. Although this seems like a letter providing a number of indisputable facts, that is a faulty assumption. For example, as an American, you are probably assuming that Jack is getting married to a woman named Pat. However, Jack could be marrying a man named Pat, because same-sex marriage is allowed in Spain. Also, an American assumption is that 10-12-09 means October 12, 2009. In Spain, however, it means December 10, 2009. Whereas Americans use month-day-year, many other countries use day-month-year. Although 9-11 is now recognized around the world as the day of the terrorist's attacks on American soil, prior to that, 9-11 was recognized in Europe as another famous day, the fall of the Berlin Wall, on November 9, 1989.

Similarly, my daughter and I recently noticed the same factual information, but employed different assumptions and came to different conclusions. Her physician recently changed her name. After learning that her physician had changed her name, my daughter remarked, *she must be so happy*. I too agreed that she must be happy. However, I later learned that my daughter assumed that she was happy because she got married and therefore changed her name. I, however, had assumed that she just got divorced, changed her name back to her maiden name, and therefore, was now happy.

Over the past several months, a local fast-food restaurant started relying on a minor, but annoying, assumption. I went through the normal steps of ordering a meal at the drive-up window. More specifically, I pulled up to the first sign, looked at the pictures of the meals, glanced at the prices below the pictures, and proceeded to order a standard meal. The price below the meal was $4.49. By the time I pulled up to the window to pay, I had already unlocked my seatbelt to pull out my wallet from my back pocket. In fact, I took a five-dollar bill from my wallet, returned the wallet to my back pocket, and had refastened my seatbelt. I had assumed that the $4.49 meal, with taxes, would be approximately $4.85. However, I was then told the price was $5.39. Rather than dispute the minor difference, I hurriedly went through the rather awkward process of again trying to quickly pull out my wallet for an additional dollar.

A week later the same problem occurred, although this time the driver behind me became impatient and tapped his horn. Two weeks later I was again charged the incorrect amount. After questioning the price, I was told that the price below the picture, $4.49, was for the *small* size meal. They had simply *assumed* that I would accept the upsized meal—*for just 49 cents more.* This form of trickery is simply known as the *bait-and-switch* technique. That is, consumers are *baited* with an appealing picture and fair price, but are then *switched* to a more expensive version. Employing such techniques and using questionable assumptions can increase profits, but at the expense of consumers.

Many jokes also take advantage of questionable assumptions. For example, suppose you meet a comedian who says, *I have two beautiful children.* A reasonable assumption is that she (or he), in fact, has two children, both of whom are beautiful. Later she remarks that indeed she has *two beautiful children and one ugly one too.*

On a more serious note, a well-known nuclear military philosophy of the United States is based on an assumption known as MAD, *Mutually Assured Destruction.* It assumes that no one will attack the United States with nuclear weapons simply because thousands of nuclear weapons will be launched in an immediate counterattack, thus ensuring destruction of the enemy who just attacked. Although MAD would deter most from launching an attack, would this assumption deter terrorists who are willing to die in carrying out a nuclear attack? Valid reasoning that uses questionable assumptions typically produces conclusions that are also questionable.

Fallacy

of

Slippery Slope

Of course people shouldn't own automatic assault rifles, but we shouldn't make it against the law. That would lead to a ban on handguns, and that would lead to a law against owning shotguns. Eventually all forms of self-defense will be taken away.

The *fallacy of slippery slope* is a flawed way of thinking, in which a person automatically assumes that supporting a particular position will produce a chain reaction of events that will inevitably lead to other undesirable consequences.

In addition, the inherent belief that this chain reaction will occur is often used as the rationale for not supporting an initial desirable position. In other words, one avoids taking the first step down a hill out of fear that the slope is slippery and will ultimately cause one to fall. Should one slip, the argument goes, sliding to the bottom is almost assured.

As in the above example, taking the first step in banning assault rifles does not automatically set in motion a chain reaction of events that ends with the banning of all guns. With no evidence that this chain reaction will occur, one would be committing the fallacy of the slippery slope. Of course, sometimes taking one step may influence the taking of additional steps; however, some people may take a few steps and realize they are going down the wrong path; consequently, they retreat to their initial position.

Taxes go up, and taxes go down. Interest rates go up, and interest rates go down. A cut doesn't necessarily lead to another cut. A step in one direction on these issues does not imply a drastic slide in that direction.

In reality, some hills are slippery, and some are not. The fallacy itself occurs only when one avoids a safe and stable hill because of the incorrect belief that it is actually slippery and dangerous. There are many truly dangerous slopes. For example, telling an initial lie often results in a cascade of additional deceptions. These deceptions, of course, are not initially intended; they are created as a means to cover the initial lie. For many alcoholics, one drink, in fact, leads to a craving for several more drinks. Using the slippery slope argument is this case is valid; consequently, the fallacy of the slippery slope is not committed.

The key to avoiding this fallacy lies in being able to accurately determine which hills are truly slippery and which ones are more like stair steps. Most people can consciously and safely navigate stair steps, whether they go up or down, in the same way that they walk up or down a slope.

The most effective way to refute an argument that uses the fallacy of slippery slope is to ask what evidence indicates that the chain reaction is likely to occur. The burden of proof is placed on the one actually using the argument of the slippery slope. If the speaker can provide solid reasons why a chain reaction will occur, the fallacy is not committed. Typically, however, the speaker will employ some weak analogy to supposedly prove his or her point. *One tiny crack in the windshield leads to larger cracks* is a statement that is generally true; however, this statement is a poor analogy for most issues that are more complex. Relying on these simplistic analogies as proof that a chain reaction will similarly occur with other issues is usually very questionable.

This fallacy is also known as the *domino effect.* It is based on the childhood game of placing hundreds of dominos in a long row and beginning the chain reaction of knocking down all of them by simply tipping the first one. Besides being called the domino effect, it is also occasionally called the *camel's nose.* Once a camel pokes its nose inside a tent, the rest of the body is sure to follow.

This fallacy is typically placed in the presumption category, simply because one is falsely presuming that one event will invariably lead to another event. As with all fallacies in this chapter, the argument is faulty because it is making assumptions that are unfounded. It overlaps with the fallacies of causation as well because, in some cases, the claim is that one event directly causes another event. It is frequently difficult to determine whether one event is just linked to another event, or is actually caused by another event.

Fallacy

of

Appeal to Ignorance

Of course I believe aliens from other planets have visited us. Not a single scientist has ever proved that they haven't. I think that is pretty strong evidence that they've been here.

The *fallacy of appeal to ignorance* is a faulty way of thinking, in which one believes that a claim is true because it has never been proven to be false, or believes a claim is false because it has never been proven to be true.

In other words, this fallacy contends that we should believe what a person says is true, not because evidence has been provided to show that it is true, but because no evidence has been provided to show that it is false. This fallacy is commonly committed immediately after someone makes a claim, and that claim is then challenged. When challenged, such individuals often respond with a statement such as, *Oh yeah, well, prove that I am wrong. You can't, so I must be right.* Despite such a self-righteous and adamant response, the response is illogical.

Such people are likely to use the same response in defending the validity of astrology, psychic abilities, or so-called supernatural powers. When supporters of such irrational and unsupported notions are asked to provide scientific proof of their existence, they often respond by saying science has failed to disprove them, and that alone is proof of their existence.

Gullible people are often guilty of this fallacy. As a result, they tend to believe in urban legends, miracle products, get-rich-quick schemes, and other scams. As Austin Cline in *Gullibility 101* notes, those who are most gullible are typically unaware of their thinking deficiencies.

> *Gullibility is an affliction, which is no respecter of class, religion, gender, or race. Gullibility is a widespread product of people's failure to properly apply reason, logic, and skepticism to a claim or idea. Unfortunately, the worse a person is at doing this, the less likely they are to realize it; indeed, they can be among those who think they do the best. As a consequence, gullibility festers and encourages the development of false, irrational, and even dangerous beliefs.*[1]

This quote reinforces a number of issues discussed in the chapter on authority. Often those who are most passionate and extreme in their opinions are also those who suffer from the highest levels of incompetence. They often vigorously voice their opinions and simplistic stereotypical conclusions. When they are challenged, they demand that you disprove them rather than take the responsibility for providing evidence for their reckless claims and opinions. Again, it is the responsibility of the one making the claim to provide evidence that the claim is true—it is not the responsibility of others to disprove the claim.

In fact, typically it is irrational to ask someone to prove that a claim is false or that something doesn't exist. Some people make claims or state opinions with such certainty and then take pride when others can't prove them wrong. The reason they cannot be proven wrong is that it is impossible to prove the nonexistence, or falsity, of many assertions, regardless of how ridiculous they are. A person may claim that it was fate or destiny that caused something to happen. When the person making the claim is asked to prove that this so-called destiny or fate had anything to do with it, no evidence is provided. Although there are some who think they are providing high-quality *evidence* by simply saying, *believe me, I just know.* When asked how they know, they often provide a circular and irrational claim such as *I know because I just know.*

The burden of proof is on the prosecution in the courtroom. That is, the prosecution must provide enough evidence to show that some standard of *proof* has been met. Most people are familiar with the normal standard in a criminal case, *beyond a reasonable doubt.* In other cases, a lower standard is sometimes acceptable—*preponderance of evidence.* Regardless of the standard, the burden is on the one who makes a charge or asserts a claim. It is not the responsibility of the defendant to prove that *he didn't do it.* That is, one is presumed innocent until proven guilty, not vice versa.

Innocent until proven guilty is a position that we endorse as being rational and logical. Shifting the burden of proof to the one arrested would mean just the opposite—*you are guilty unless you can prove your innocence.* If you can't prove your innocence, you are therefore considered guilty. This is a ludicrous scenario because there are often dozens of initial suspects in serious criminal cases. Should these suspects simply be charged with a crime and then also be responsible for proving that they are innocent? This would violate the rules of logic and the Bill of Rights, as well as increase the prison population by 1,000%. In all probability, such an irrational system would lead to the conviction of several individuals for the same crime.

Although the prosecution cannot shift the burden of proof to the defendant, the defendant cannot logically employ this fallacy either. For example, suppose the prosecution presents overwhelming evidence of guilt. Suppose further that the defendant is found guilty. However, the defendant declares that a *voice in his head from the devil* told him to commit the horrendous crime. He declares he should not be sent to prison and that the prosecution must now prove that he did *not* hear the voices he is blaming for the crime. In this case, the burden of proof is on the defendant to prove that he is insane, not the other way around.

As was mentioned previously, it is possible to prove that something exists; however, to prove the nonexistence of something is nearly impossible in many cases. *Proving,* for example, that the so-called *Bigfoot,* the missing link between man and the ape, does *not* exist would require that every square foot on the globe be searched. *Proving,* for example, that the *Loch Ness Monster* does *not* exist would require that the entire lake be drained.

This rational position is the view of the *skeptic.* Despite its negative connotation, being a skeptic is an extremely logical and scientific way of looking at the world. In the absence of evidence, skeptics and scientists simply say they don't know whether something exists because there is no evidence allowing them to take any other position.

A brief review of the most fundamental concepts in critical thinking must be understood in order to understand why the *you can't prove me wrong* position is illogical. Critical thinking is based on the well-established principle that something should not be accepted unless supporting evidence is provided. That is, saying something is true is not evidence for its truth.

For example, those who entirely reject the scientific notion of *evolution* are taking a position that is completely contrary to overwhelming evidence. In contrast with *skepticism, cynicism* is irrational and illogical. Some cynics continue to claim that the world is flat, that the holocaust didn't occur, that man never went to the moon, and that the federal government caused the twin towers of the World Trade Center collapse with carefully concealed dynamite and detonations. Such conspiracy theories feed on paranoia and irrational cynicism.

In the field of medicine, respected physicians often acknowledge that the effectiveness of many treatments is unknown, simply because adequate medical research has not been done. They are saying that, at present, the evidence neither supports nor rejects the effectiveness of a particular treatment. Although they don't know whether a treatment will work or be effective, they are open to the idea of trying to find evidence that either supports or refutes effectiveness. Again, this is a logical and rational response to advancing science and medicine.

So, the fallacy of appeal to ignorance says we cannot assume something is true because it cannot be proven as false. However, another part of this fallacy says that we cannot assume that something is false because it cannot be proven as true. Going back to the previous example, we cannot say that *Bigfoot* or the *Loch Ness Monster* do *not* exist simply because no good evidence has been found for their existence. We cannot, for example, say there is no life on other planets simply because we currently have no proof of life elsewhere. As William Cowper said, *absence of proof is not proof of absence.*[2]

In other words, the lack of evidence does not prove that it doesn't exist. There are times, however, when one can reasonably conclude that something doesn't exist if an adequate search has been completed. For example, it is not possible to say with *absolute certainty* that Iraq does *not* have weapons of mass destruction because every square inch of the country has not been searched. However, given the scope and length of the search, it is certainly reasonable to assume that the weapons probably do not exist because they would have been found by now if they existed.

So technically, this fallacy suggests that we should neither accept nor reject a claim without evidence. In practical situations, however, we are often forced to make a choice, even when there is no evidence. What should we do? Logically, we should normally deny the claim. As in law, the default position is *not guilty.* If someone claims that unicorns actually exist, but provides no evidence for or against that claim, we should assume it is *not true.* If this were not the case, we would be obligated to accept every claim of every person, regardless of how preposterous such a claim might be. When in doubt, we should not accept. Thinking something is true, when its not, is worse than not knowing at all.

Fallacy

of

False Charge of a Fallacy

You are telling us that we should be vegetarians, but at the same time you are wearing a leather jacket. That makes no sense!

The *fallacy of a false charge of a fallacy* is a faulty way of thinking, in which one accuses another of committing a fallacy when no fallacy was actually committed. Consequently, the person who made the false charge is actually the one guilty of committing the fallacy.

People who have very strong beliefs or convictions about various issues often commit this fallacy. That is, they are so certain of their position that they fail to consider that they may have reached their conclusion through either incorrect information or faulty reasoning. When such individuals' arguments are refuted with contradictory evidence or logical reasoning, rather than refute the criticism, they often counterattack by accusing the critic of reasoning in a fallacious way or relying on biased information. Typically, though, they are unable to explain why the critic's reasoning is fallacious.

In the example above, one individual is accusing another of committing a fallacy. However, the one wearing the leather jacket is not necessarily guilty of a fallacy; consequently, the person who made the false charge is actually the one guilty of a fallacy. Generally speaking, a vegetarian is typically defined as one who does not *eat* meat (red meat, poultry, or fish). This does not prevent a vegetarian from *wearing* leather made from animals. So, if this definition is used, the critic actually committed the fallacy, not the person wearing the leather jacket.

We are defining a fallacy as an error in reasoning or using data that are incorrect. Even the first sentence in this paragraph might tempt some people to commit a false charge fallacy. That is, some readers might charge that a grammatical error was committed by writing the statement: *the data* are *incorrect*. In fact, however, *data* is plural and *datum* is singular; consequently, writing that *data are* is grammatically correct. Suggesting that the writer made a grammatical error in this case would be a false charge. On the other hand, there are probably a few other cases in this book where there are genuine grammatical errors.

Although this is a rather trivial grammatical example, there are many other religious, economic, and *so-called political truths* that people *know* to be true. And many people articulate these matter-of-fact statements on a regular basis; however, careful research reveals that most of these so-called truisms are myths. When one refutes the argument of a person who is repeating one of these myths, often the one spreading the myth simply says his opponent is uniformed or reasoning in a faulty way. In fact, however, it is the one spreading these myths who is actually committing a fallacy.

Without going into detail, there are many uninformed individuals who fall victim to smear campaigns of the political parties. That is, much of what people believe about the Democratic and Republican parties or conservative or liberal philosophies is simply not true. Propaganda that reinforces these misconceptions is fed to voters almost daily.

Some news organizations and publications strongly favor certain political parties. In fact, as mentioned elsewhere, some publications openly admit their general political leanings. Unfortunately, there are millions of people who rely predominately on one or two of these biased sources for their news.

Political talk radio shows, for example, attract listeners who initially have strong political leanings. Although not stated up front, the goal of the show is to reinforce the views that these listeners already have. As a result, their views become even more intense and more extreme. People become what they are fed. Rather than develop a healthy diet in which a wide array of political and social perspectives are consumed and digested, many people fall into the habit of eating the junk food of sound bites, headlines, and political propaganda. People often simply seek information to confirm their pre-existing opinions. Psychologically, it is very comforting to have opinions reinforced, even if they are wrong.

On a lighter note, there is an often-told joke about two engineers who are trying to figure out how to determine the height of a flagpole. They are frustrated because each time their tape measure extends more than several feet up the pole, it bends over and falls to the ground. A few minutes later a blonde (or brunette, redhead, etc.) walks by and sees their frustration. Note that the blonde could be a man or a woman. The blonde simply removes a couple screws at the base of the pole, lays the flagpole on the ground, and easily measures it with the tape measure. After the blonde leaves, the engineers remark, *it is just like a blonde to measure the length, when the height is needed.* In reality, the blonde's method of determining the height was perfectly sound. So, in criticizing the blonde, the two engineers committed a false charge of a fallacy.

Many people charge politicians of a fallacy when they change their vote or opinion on some issue. While the politicians would be guilty of a fallacy if they changed their opinions for no good reason, changing one's mind does not necessarily constitute a fallacy. For example, DNA samples that were gathered many years ago at crime scenes are now being reanalyzed with new DNA-testing techniques. In some cases, the new analysis shows that a crime for which an individual was convicted and jailed was actually committed by someone else. Given this new evidence, judges sometimes overturn a previous conviction. Such a reversal does not commit a fallacy because the change was based upon new information that was not available at the time of the original trial.

When people commit fallacies, it is perfectly appropriate to point them out. But, anyone who charges another with using a fallacy must be certain that the original argument was actually a fallacy. Otherwise, the critic actually becomes guilty of a fallacy for falsely accusing another of using fallacious arguments.

In summary, this fallacy falls into the presumption chapter because unwarranted assumptions are made. Whether it is assuming that vegetarians cannot wear leather or that a pole's length is different from its height, there is some statement or idea in the claim that is questionable.

Fallacy

of

Apriorism

> *There is no way that my friend is a killer. I just won't believe it. I know more than a dozen people said they saw her repeatedly stab her husband. I know her blood and fingerprints are on the knife, but I know she didn't do it. I don't care how much evidence you have, you can't convince me that my friend did it.*

The *fallacy of apriorism* is a faulty way of thinking in which one reaches a conclusion before (or prior to) knowing the facts, rather than looking at the evidence and then drawing a conclusion.

Imbedded within *apriorism* is the word *prior,* which identifies the key aspect of this fallacy and simply means *before* or *previous.* An *ism* at the end of a word typically takes the meaning of the word and expands it to include a belief system or philosophy. For example, *extremism* is simply a philosophy of endorsing radical (extreme) ideas or behaviors. Similarly, *conservat**ism*** is a belief of retaining (or conserving) the status quo or traditional beliefs rather than accepting ones that invite change. Consequently, *a prior ism* is a statement or unwavering belief that is made prior to examining the evidence.

This fallacy is sometimes called *invincible ignorance* because the one committing the fallacy will remain ignorant of the truth, regardless of how much strong evidence is provided. In the example above, the individual's certainty of her friend's innocence is *invincible* in that it cannot be shaken or beaten even in the face of overwhelming evidence. Having faith, or the hope of a friend's innocence, is far different from delusional thinking in which one is blinded to facts.

My wife and I suspected that this *fallacy of apriorism* occurred the first time we took our son to see the orthodontist. Our goal for this initial appointment was to determine whether our son even needed braces, because his teeth looked nearly perfect. Upon arriving at the office, I had assumed that the orthodontist would first complete the dental examination, and then, if necessary, we would discuss our insurance coverage and a payment plan. Rather than meeting first with the orthodontist, we met his assistant, who eagerly verified our dental insurance and recorded our work history, and I suspect even checked our credit scores. After these preliminaries, the assistant implied that it was a forgone conclusion that he would need braces. This conclusion came before our son opened his mouth for an examination. In other words, our experience seemed to match with the stereotypical notion that *if the parents have jobs, the child needs braces.*

Unlike the scientist who carefully takes various bits of information and then tries to construct a coherent and unified theory or principle, those guilty of this fallacy do just the opposite. That is, they begin with a preconceived idea or belief, and then filter all evidence so that it supports the initial belief. Evidence that contradicts the theory is refuted or dismissed, often unconsciously. This act of dismissing information that runs counter to a belief is further perpetuated by the act of searching for only information that supports one's theories. As mentioned in another chapter, social scientists call this perceptual error *confirmation bias.* In other words, people have a bias in that they search only for information that confirms their preconceived ideas.

Religious fundamentalism and political extremism rely on the fallacy of apriorism on a regular basis. For example, there are some who *know* with certainty that the world is just 6,000 years old because they believe every word in the Bible should be taken literally. And, they have been *told* that the Bible *says* the earth is just thousands of years old. (Actually the Bible includes no such claim). Despite scientific proof of fossils millions of years old, some dismiss all scientific evidence in order to hold onto their cherished, but irrational beliefs.

In the same way, some extremists view all ultraconservative ideas in a positive light and all liberal ideas in a negative light. Far-left liberals have a corresponding and opposing view. Such preconceived ideologies are so strong in certain political circles that powerful and legitimate evidence becomes almost irrelevant to decision-making.

The overwhelming power of the preconceived conclusion has been investigated in both actual and laboratory situations. One of the best-documented cases involved the ill-fated decision to launch Challenger, the space shuttle that exploded just 72 seconds after takeoff from Cape Kennedy in 1986. The cause of the explosion, a failure of the O-rings, did not come as a shock to some at NASA. In fact, engineers had repeatedly expressed their worries about the O-rings just days before the launch. Within hours of the explosion, major news networks were already reporting on the O-ring situation. An analysis of the decision-making process concluded that NASA suffered from *groupthink.* One symptom of *groupthink* involves the tendency to disregard information that contradicts the general group norms of striving for agreement. That is, dissent or disagreement is often suppressed. Despite repeated concerns expressed by several engineers about the O-ring problem, NASA decision-makers seemingly had already made a prior decision to launch.[3]

The attack on Pearl Harbor is another case in which significant evidence was ignored. Despite warnings from Washington in 1945 that an attack from Japan was imminent, American military commanders did nothing to protect the navy fleet from attack in Pearl Harbor.[4] Their preconceived belief that they were not susceptible to an attack prevented them from seeing that an attack was forthcoming.

Last, one of the most thorough congressional investigations in history concluded that the CIA's assertions that Iraq had weapons of mass destruction *were unreasonable and largely unsupported by the available intelligence.*[5] This assertion led to the invasion of Iraq, an invasion that sacrificed thousands of American lives, more than 100,000 Iraqi lives, and cost the United States more than a trillion dollars.

Laboratory findings shed some light on why these decisions from apriorism occur. Consider the following study as reported by author Steve Denning.

> *In 1979, a psychologist named Charles Lord and his colleagues at Stanford University published their classic research on what happens when people are presented with arguments that are at odds with what they currently believe. Lord's team selected twenty-four proponents and twenty-four opponents of capital punishment. They showed them studies that confirmed the penalty's deterrence as well as other studies that refuted it. What happened? The proponents of capital punishment interpreted the studies as supporting capital punishment, while the opponents of capital punishment concluded that the evidence refuted the approach. Both proponents and opponents found clever ways to reinterpret or set aside any contrary evidence so as to confirm their original positions.*[6]

In other words, for both groups, the *evidence* itself had *no effect on their views.* Put another way, people previously made up their mind and no amount of evidence was going to change that view. In fact, even in the face of massive evidence against their position, rather than moving to a more moderate position, they became even more polarized and harden in their initial opinion. They absorbed the information that supported their belief and dismissed any information that opposed their belief.

Studies suggest that what held true in laboratory studies in 1979 also held true in the 2004 presidential election. Consider the following research.

> *But why aren't we more willing to reconsider our positions in the face of serious factual evidence that should at least give us pause? Aren't we thinking at all? Apparently not, according to a recent study by psychologist Drew Westen and his team at Emory University. The team conducted Functional Magnetic Resonance Imaging (MRI) brain scans on fifteen "strong Republicans" and fifteen "strong Democrats" in the course of the 2004 presidential campaign while they were reviewing blatantly self-contradictory statements by the two candidates, George W. Bush and John Kerry. As we would expect from earlier studies of the confirmation bias, the Democrats found ways to reconcile Kerry's inconsistencies and became even more strongly Democrat, while the Republicans had no difficulty explaining away George W. Bush's self-contradictions so as to become even more fervently Republican.*[7]

This *Fallacy of Apriorism* helps explain why political attitudes and voting patterns are so difficult to change. After a typical presidential debate, the question posed to voters across the country is: *Who Won?* For the majority of people, the answer to that question is determined long before the debate even begins. Overwhelmingly, registered Republicans will say that the Republican candidate won. In contrast, Democrats will say overwhelmingly that the Democratic nominee won. The vast majority of the voters typically have their minds made up prior to even knowing who the nominee is—and regardless of the positions the nominees articulate. Many people simply assume that they will vote for whomever *their* party nominates.

In fact, a significant percentage of people vote a straight party ticket, but cannot even explain their party's position on a host of relevant issues. *I've always been a Republican so I'll vote for the Republican. I've always been a Democrat so I'll vote for the Democrat.* For the vast majority of voters, the debates themselves have virtually no effect on their votes. An analysis published by *Time Magazine* (October 2008) provides some insightful details. Voters are roughly divided into four equal groups: about 23% do not vote, about 26% vote straight republican, about 27% vote straight democratic, and about 24% are swing voters. A swing voter is defined as one who is not firmly committed to a candidate or party.[8] In other words, the debates and the candidates themselves have little to do with how about 75% of the electorate votes (or doesn't vote).

Simply ignoring the individual candidate and voting for a particular party's candidate is not necessarily illogical. It becomes illogical, however, when the voters do not know the position the party holds. On average, another study reported in *Time Magazine* found that about 28% of the voters unknowingly vote for the wrong candidate or party. That is, they are so ill informed about the party or candidate's position that they end up voting for the candidate who does not share their views.[9]

Such mindless decision-making negates the goal of critical thinking, the act of thinking thoughtfully. The problem with making preconceived decisions is that a great deal of relevant evidence on the issues is simply ignored. As was mentioned previously, once people have a belief engrained on their brains, they tend to search for evidence to support their beliefs and refute evidence that contradicts their beliefs.

After prescribing new drugs for their patients, it is standard practice for most physicians to warn them about potential side effects of the drugs. However, given this warning, many patients become hypersensitive to the slightest changes in how they feel. In other words, if they have a preconceived idea that they will have side effects, their brains actively seek out such information to support the preconceived idea.

As implied above, for people who have extremely strong political, religious, or ideological beliefs, evidence and logical reasoning are *not* powerful tools of persuasion. This is often why so much advertising and campaigning focuses on appealing to emotions.

This *confirmation bias* literally blinds many people to the actual evidence. This is why evidence must be thoroughly considered before making a decision. Once we make an initial decision, it is very difficult for us to change our mind in midstream. Once in the middle of the stream, we look for evidence supporting our opinion. Rarely do we search for evidence that contradicts or refutes our initial opinion. Critical thinkers, however, are willing to do just that—even though doing so might be uncomfortable. Although it is tempting to make a decision without ample evidence, critical thinking demands otherwise.

Perhaps the best way to sum up this fallacy is to look at two popular quotes. *I will believe it when I see it* simply warns us not to believe a claim without evidence. However, the famous slip of the tongue from social psychologist Thane Pittman is far more insightful—*I will see it* ***when*** *I believe it.*[10] This quote simply reminds us that if people believe something to be true, they will suddenly find all kinds of support for the belief. Contradictory evidence, however, is ignored.

To a large extent, prior beliefs influence our perceptions much in the same way that binoculars artificially magnify certain elements in our world. However, these beliefs, like binoculars, often prevent us from seeing what is right next to us.

Fallacy of Moving the Goalposts

Jack: *Clearly the war in Iraq is a success.*
Sally: *In what way?*

Jack: *Well, we accomplished the mission to destroy Iraq's weapons of mass destruction.*
Sally: *There were no weapons of mass destruction.*

Jack: *Oh. I mean our goal was to kill Saddam Hussein, who ordered the 9-11 terrorist attacks.*
Sally: *Saddam Hussein had nothing to do with the 9-11 attacks. That was Osama Bin Laden.*

Jack: *Oh. I mean the mission was to capture the brutal dictator, Saddam Hussein.*
Sally: *First, you said the mission was about weapons of mass destruction. Then you said the mission was about the 9-11 terrorists. Now you say the mission was about Saddam Hussein.*

Jack: *We captured a brutal dictator. So, the mission was a success.*
Sally: *Yes, we did capture a brutal dictator, but that was not the primary mission. If the goal was to capture a brutal dictator, there are at least ten other dictators even more brutal than Saddam Hussein.*

The fallacy of moving the goalposts is a faulty way of thinking in which one keeps changing the reasons or standards for accepting or rejecting a claim. Whether one is making an argument or playing a game, the rules of the game are agreed to in advance of the game itself. In the case of making an argument, the rules of the game are simply the standard principles of logic.

In the example above, the standard for *mission accomplished* was agreed to long before the war started. The standard for success was the destruction of Iraq's weapons of mass destruction. From a logical point of view, the mission failed because this standard was not met. In fact, it was impossible to accomplish because the weapons didn't exist.

As most people know, there are goalposts at the ends of a football field, and when a team is able to kick the ball through the posts, the play is successful. The posts are stationary, and the rules are very clear—points are scored if the ball goes between the posts (success)—no points are scored if the ball is short or wide of the posts (failure).

After your team successfully kicks the ball through the upright goals, assume that the referee says the play needs to be redone for some reason. However, before your team kicks the ball again, the goalposts are moved ten yards beyond where they were previously. Obviously your team protests the new standard for success, but your protest is rejected. Although the goals are ten yards farther away, your team is still able to kick the ball through the upright goals. Again, the referee says the play must be replayed, and again the referee moves the goalposts another ten yards farther away. Regardless of how great your kicker is, at some point he will be unable to kick the ball far enough. As soon as he fails, the referee will announce the failure as legitimate and it becomes the official outcome of the play.

So, in the example above, failure is assured. On the other hand, the fallacy sometimes works the opposite way as well. In other words, the goalposts can be moved closer until success is guaranteed.

In looking at the example about the war, Sally is amazed at Jack's inability to follow the basic rules of logic. Jack, like millions of other people, routinely commits this fallacy. People sometime cling so passionately to a belief that they will run over any roadblocks that stand in their way. In fact, they often don't even see the roadblocks—or the errors in reasoning. That is, they genuinely convince themselves that they have used valid reasoning to arrive at a rational position. In other cases, persuaders will use this fallacy consciously and simply hope that others will not see the reasoning flaw. The first reason for the war was unfounded, so Jack moved the goalposts closer, but his second attempt to claim the mission of the war was to capture the 9-11 terrorists also turned out to be false. In other words, he missed the field goal twice. Once the goals were moved again by asserting that the real mission for the war was to capture a brutal dictator, he was finally able to kick an argument through the goalposts and artificially declare victory.

Every military action, including the war in Iraq, produces both positive and negative outcomes. But, regardless of the end result, the publicly stated and primary reason for the war was to destroy Iraq's weapons of mass destruction. Once it was determined that Iraq had no such weapons, the primary foundation on which the war was waged vanished. Consequently, the primary argument for starting the war turned out to be a bad argument.

Once it was acknowledged that the original mission for the war could not be accomplished, some might argue that withdrawing from Iraq would be the only logical option remaining. There is merit to this argument. However, despite the fact that the war was started under false assumptions, some might argue that there are legitimate reasons that the war should continue as a way of preventing our first mistake from becoming even more disastrous. That is, some have argued that continuing the war would minimize the damage that had already occurred, whereas others have argued that continuing the war would cause even more harm.

Suppose, for example, you decide to buy $10,000 worth of stock in some company, but soon after your purchase the company announces it has financial problems. The stock falls dramatically. Your initial decision to buy the stock appears to be a mistake. Should you instantly sell the stock, admit your mistake, and just accept your loss? Or, should you keep the stock and hope that one day your investment will pay off? Or, is there a possibility that your initial mistake will become even a bigger mistake if you hold on to a stock that continues to drop? Should we cut our losses and get out? In the case of Iraq, the previous White House policy appeared to be that this was an investment that would eventually pay off. Was it a good investment, or will it turn out to be a case of *throwing more good money after bad money?*

This very example brings us to a closely related flaw known as the *fallacy of sunk cost.* This fallacy reasons that we must continue in the same direction, regardless of the costs, simply because we have already sunk a great deal of money into our effort. If we do not continue in the same direction, the argument goes, we will have lost everything. If there is a high probability of success with a slightly greater investment than was initially expected, moving forward may be logical. On the other hand, the size of one's initial investment is not normally a good reason for continuing down the same path. Losing a million dollars in an investment is not a good reason to invest another million dollars. Although it is true that the initial investment will have been lost, investing further in a plan that has little hope for success is unreasonable.

Of course, if one continues to invest in an effort regardless of cost, most efforts can be *seemingly successful.* For example, suppose you invest $20,000 to drill a well to supply your home with water. The driller, after exhausting your investment and still not hitting water, persuades you that perhaps another $10,000 to drill 200 feet deeper is needed. Later, you are again persuaded to pay another

$10,000 to go still 200 feet deeper. After this investment, again no water is found. If one drills deep enough, sooner or later, water can be found virtually anywhere. The question, of course, is whether the costs sunk into the venture exceeded the potential benefits.

Politicians in favor of continuing the war will suggest that standards of progress have been met, and sinking more money into the effort is worth it. However, many opposed to the war will counter that such politicians are moving the goalposts, in that they keep lowering the standards that define *success.*

This fallacy occurs with important policy issues, but it also occurs frequently with more mundane issues. For example, in playing a game such as table tennis, both you and your partner agree that the winner will be the one who wins two out of three games, with each game played to 15 points. After apparently losing the first game by a score of 15-13, you contend that a *true* game is played to 21. Later, after you have lost two out of three games, you argue that it should be the best of five games. In order to prevent a loss, you keep changing the rules.

Most politicians will go to great lengths to avoid having to say they made a mistake. It is a sign of weakness, they reason, because making a mistake in the past is an indication that they might therefore make one in the future. Many such people will employ this fallacy of moving the goalposts to avoid losing, looking foolish, or appearing incompetent.

In fact, in 2005, then-president George W. Bush admitted that he would have gone to war with Iraq *even if he knew then what he knows now*. That is, even if he knew there were no weapons of mass destruction, he *still* would have initiated military action.[11]

This directly contradicts the stated rationale for going to war—to destroy the weapons of mass destruction. The fact that some good came from the war (the removal of a brutal dictator) does not allow one to logically claim success.

Such illogical reasoning might be politically wise; however, it unquestionably violates the rules of logic. So strong is this desire to avoid admitting a mistake that even President Bush's opponent in the 2004 election, Senator John Kerry, also stated that even *if he had known then what he knows now,* he *still* would have voted to authorize force against Iraq in 2002.[12]

Both candidates probably reasoned that the public would see them as inconsistent or weak if they changed their initial positions. As noted in a previous fallacy, it is illogical to change one's mind for no apparent reason; however, when there is new evidence, changing one's mind is entirely logical. It might not be popular, but it would be logical.

Both Bush and Kerry directly violated the rules of logic when they asserted that they would have ordered the war even if they had known that Iraq did not have weapons of mass destruction. The original argument for the war was based squarely on a mission to eliminate such weapons. Without the weapons, the argument for the war also evaporated. That is, there could be no success if there was no mission.

Like one who changes the rules for what constitutes *success*, many so-called psychics or astrologers employ many excuses, include moving the goalposts, to avoid looking incompetent when they have clearly failed. Even when such individuals have agreed to have their so-called *powers* tested under scientific conditions—conditions that they have agreed to in advance—they explain away their failures with paltry excuses. Common moving-the-goalpost excuses include the claim that *nonbelievers* in the room created negative mental energy that interfered with their psychic abilities. *Being too tired* and *on some days my powers are weak* are other common excuses. Another common way that strong believers move the goalposts after a psychic has failed is to claim that this particular psychic was not a *true* psychic.

One way that people *move the goalposts* is by relying on another fallacy known as *no true Scotsman.* In this case, the goalposts of the arguments are moved by adding exceptions or conditions to make the original claim appear to be true even when it is not. The name of this fallacy stems from the claim that *Scotsmen never put sugar on their food.* When shown a Scotsman who uses sugar, the claim is maintained by saying no *true* Scotsman puts sugar on his food. Or, as mentioned above, when a psychic is unsuccessful, the claim is that she was not a *true* psychic. Similarly, suppose an individual makes a claim that *all scientists believe in evolution.* When a scientist steps forward and says that he doesn't believe in evolution, the one who made the original claim can retort by saying all *true* scientists believe. In applying this fallacy, one could replace *true* with such words as *honest, credible, moral, decent, respected, etc.*

Politicians employ this fallacy when they make a promise or state a position, but later find they need to change their position. Rather than admit that they are changing their position or breaking a promise, they add conditions or clauses to the original claim.

For example, the official White House position in 2008 was that there would be *no negotiations* with Iran. However, in the summer 2008, Undersecretary of State

William Burns was sent to Geneva for talks with Iran. To explain this apparent contradiction, the new rhetoric was that there would be no *separate* negotiations with Iran.[13]

Similarly, in 2008, Presidential candidate Barack Obama employed this approach when he added a condition to his original position of opposing offshore drilling for oil. Later he asserted that he was opposed to offshore drilling *unless* it was part of a comprehensive energy program.[14]

Presidential administrations from both major political parties routinely rely on this technique to give the impression of taking a firm stance, while the ground below their feet is slowly shifting to a new position. Because the public is often wary of changing positions, even in light of new conditions, one cannot be surprised that policymakers frequently rely on this verbal technique.

Extremists, especially those who have radical religious, social, or political views, and those who believe in paranormal abilities often are poor critical thinkers, despite their passionate convictions. As such, they often go to great lengths to defend their beliefs. As mentioned previously, those who suffer from poor thinking are also the *least* likely to realize the irrationality of their thinking. *Moving the goalposts* allows them to claim their opinions are supported when the evidence clearly indicates the opposite.

Fallacy

of

Contradictory Premises

> *My husband doesn't like most of the perfumes that I wear. So I just bought myself a bottle of odorless perfume. I love the perfume, and so does my husband.*

The *fallacy of contradictory premises* is a faulty way of thinking, in which one believes that a logical conclusion can be reached even when the grounds on which the argument rests cancel out each other or cannot both be true at the same time.

In order to understand this fallacy, a quick review of the parts of an argument is needed. As mentioned in Chapter One, an argument has premises and a conclusion. Reasoning is applied to the premises to produce a conclusion. For example, premise *one* might state that Iowa has about three million people, and premise *two* might state that Arizona has six million people. Through reasoning, the conclusion that can be reached is that Arizona has more people than Iowa.

In the *fallacy of contradictory premises,* one sets up an argument that has two premises but upon closer inspection, the two premises cancel out each other—they are contradictory. If the two premises cannot both exist at the same time, no genuine argument is created, and no conclusion can be drawn. That is, the fallacy of contradictory premises is committed.

In the example above, one premise is that all perfume has an odor. Another premise is that the perfume that the wife bought is odorless. At this point, because premise *one* and premise *two* cannot both be true, no argument can be constructed.

Another example that is frequently mentioned in various sources is the age-old paradox about God and an immoveable object. The following question is posed: *Can God, **a being that can do all things**, make a stone so heavy **that he cannot lift it?*** The first premise is that God can do all things. The second premise is that there exists something that cannot be done (a stone so heavy that it cannot be moved). If God can do all things, but can also make a stone so heavy that he can't lift it, in reality he cannot do all things. The two premises directly contradict each other and cannot both be true. Therefore, this argument is a fallacy.

(Actually it is not even a bad argument; technically it does not even qualify as an argument).

Usually one can easily recognize such fallacies. My students, for example, instantly recognize the conflicting position I place them in when I discuss my attendance requirement. Teasingly, I tell students that absences due to personal emergencies will be excused—as long as I am notified two weeks in advance of the emergency. Again, emergencies are such that they cannot be known in advance. Otherwise they wouldn't be emergencies.

Another often-cited paradox occurs when one premise states that cats always land on their feet. Suppose we accept this premise as true, and we also accept the premise that, if buttered toast falls, it always lands buttered side down. Given these two premises, what would happen if you were to take a cat and strap a piece of toast, buttered-side-up, to its back—and then throw the cat out the window?

Another philosophical dilemma occurs when I contend that everything I say is true. Then, I proceed to say that my previous statement is untrue. If I always speak the truth, then the previous true statement must be false.

Similarly, Groucho Marx said, *I refuse to join any club that would have me as a member. Please accept my resignation.*[15] Many jokes, such as this one, rely on inconsistency. One premise is that I want to join a club, but the second premise is that I would not join any club that has such low standards that it would admit me.

Politics is laced with contradictory premises. In Arizona in 2008, the House Appropriations Committee approved a measure that would bar public schools from teaching anything that is antidemocratic or that opposes Western civilization. This measure would do the opposite of what it is intended to do. It is a contradictory premise fallacy because there are two premises in the claim that cannot both be true—they are contradictory. One fundamental premise of democracy is the freedom to criticize its government and the democratic process. In fact, free speech is the cornerstone of democracy. But, the underlying premise of the bill is that it will become illegal to criticize democracy.[16]

Sometimes the contradictory premises are blatantly obvious. For example, suppose there are two young women who are identical twins—they look and dress the same. One twin believes herself to be ravishingly beautiful. But, she criticizes her sister as being grotesquely ugly. Again, this is an argument that cannot exist because it would require accepting two premises that contradict. If two people are biologically identical and look precisely the same, one cannot be ugly and the other one beautiful.

Likewise, if I ask my students, *What is a rhetorical question?* and no one answers or raises a hand, what can I conclude? I really cannot conclude anything, because the very question I asked was a rhetorical question, which is a type of question that seeks no response or answer. It creates a contradictory premise situation because, if someone raises his hand and correctly explains the answer, he is obviously knowledgeable. If, on the other hand, no one responds, this lack of response is also assumed to demonstrate an understanding of a rhetorical question.

An expert who says that experts cannot be believed also creates a paradox. If we believe this expert, then we cannot believe his statement. But if we believe his statement, then we cannot believe him. Similarly, it could be argued that a billboard with the words, *Keep Your Eyes on the Road,* contradicts itself. That is, in order to read this warning, we must ignore the warning, and actually take our eyes off the road to read the sign.

As mentioned in a previous fallacy, one premise of democracy is that we govern ourselves by majority rule. On the ballot in Arizona in 2008 was a referendum known as: *Majority Rules—Let the People Decide: Proposition 105.* Although this ballot has an appealing name, it would remove this basic premise of letting the majority decide. This proposition would allow just a few people to dominate the vast majority. To be more specific, Proposition 105 would require that all future ballots that involved any spending be approved by the majority of all *registered voters, not simply by a majority of those who actually voted.*[17]

In other words, every person who did not vote would automatically be counted as a NO vote. Suppose that 50 percent of the people voted, and further suppose that ***99%*** of the people voted in ***favor*** of the proposition. The 1% of those who opposed the measure would defeat it. That is, the 99% who voted in favor of the proposition would be defeated by the 1% who opposed it (along with the 50% that did not vote).

The voters rejected this ballot in 2008 by a 2-1 margin. Had it passed, the chances of any future ballot passing would have been virtually zero. There have been scores of ballots in Arizona since 1992, and a fair number of them have passed with a simple majority. Had this proposed requirement been in place, not a single ballot would have passed.[18]

Given the appealing wording of Proposition 105, *Majority Rules,* many people unknowingly used the principle of *majority rules* to vote against the principle of *majority rules.* The premise hidden in Proposition of 105 directly contradicted an accepted democratic principle. Consequently, the *Fallacy of Contradictory Premises* occurred.

This *fallacy of contradictory premises* is also known as the *fallacy of conflicting conditions.* Sometimes it is also called a *logical paradox.*

Fallacy

of

Either-Or

> *Look, it is very simple. Either you are* ***for*** *the terrorists or* ***against*** *the terrorists. Either you* ***support*** *the war or* ***oppose*** *it, which is it?*

The *fallacy of either-or* is a flawed way of thinking, in which one claims that there are only two possible choices, when, in fact, there may be more than two.

This fallacy occurs frequently and is often committed by those who see the world in simplistic black and white ways. That is, they divide issues into good and bad, right and wrong, effective and ineffective. They don't realize that most issues can be analyzed from various intermediate and rational perspectives. Although those who lack the sophistication to see these other perspectives often commit this fallacy, those in power often commit it as well.

Many politicians exploit the fact that public officials who take clear and absolute stands on various issues easily impress many voters. Even when an outcome is entirely unknowable, a politician will imply that ***either*** *you vote for me (which is a guarantee of prosperity)* ***or*** *vote for my opponent (which is a guarantee for disaster).* Both the either-or nature of their advertising and the absolute tone of their guarantees violate the principles of logic. However, because some voters do not understand or value the importance of logic and rationality, politicians can use emotional or irrational appeals with great success.

Although those who take extreme either-or positions on such issues can often attract votes, these positions often turn out to be unrealistic because of changing circumstances. *I will never increase taxes* is a statement that has been articulated by many politicians; however, in the end, such positions are often modified. Another common political promise is to increase funding for education. This promise, as well, often goes unfulfilled. In fact, vowing never to increase taxes, yet promising to increase funding for education, are frequently competing priorities.

Religious extremism often leads to black and white thinking. *Either you believe or don't believe* and *either you will go to heaven or to hell* are two statements that extremists often make. In making these statements, such people often let their beliefs overpower their intellectual reasoning. That is, their rational powers of reasoning might suggest that there are more than two choices; however, they fear that questioning their beliefs through reasoning is somehow disloyal or morally wrong.

The abortion debate is frequently, and incorrectly, presented as if there were only two options: pro-choice and pro-life. Even the terms used to describe these two positions incite extremism. If this issue is placed on a continuum, it could be argued that there are hundreds of intermediate positions. There are those who hold absolute and adamant views that the willful termination of any pregnancy, *for any reason,* is murder. On the other hand, extremists on the other end of the continuum suggest that abortion should be promoted as a standard birth control method. Most people don't hold such extreme views.

In fact, many people hold views that seem to be pro-choice and pro-life at the same time. Even most of those who are strongly pro-choice are not pro-abortion. This is not a contradiction. That is, they do believe in a woman's right to choose, but they strongly believe that everything should be done to prevent abortion. They strongly believe in family planning and birth control education. They also strongly believe in promoting adoption rather than abortion. Like most people, those who hold these views, cannot be placed in one either-or category.

In presidential elections, the notion that we must vote for either the Democratic or Republican nominee is not entirely true. In reality, one can simply not vote, vote for an independent candidate, or even write in a candidate's name. Similarly, one might believe that many organizations support one candidate and oppose the other. In reality, some organizations contribute to both political candidates. Why would they do this? Doesn't this basically cancel out their contributions? Although not explicitly acknowledged, candidates find it difficult to ignore those who have contributed heavily to their campaigns. By donating to both candidates, lobbyists cover all their bases and are virtually guaranteed some influence on or access to the winning candidate.

In speaking about presidential candidates, Barack Obama is one who clearly does not *fit* into the normal either-or categories. Although such candidates do not fill out a *job application form* identifying their race, how would such an individual mark off the *race* box? Although most voters perceive Barack as black, he is actually biracial, having a black father and a white mother. Similarly, Tiger Woods, a well-known professional golfer, is part white, part black, part Native American, and part Asian.[19]

On a more practical level, some individuals allow themselves to be cornered into giving either-or answers in legal situations. Although an attorney might ask a question and demand a *yes* or *no* answer, there are often many other answers that a witness can provide. *I don't know, I don't remember, maybe,* or *I refuse to answer that question based upon legal advice* might seem like ways of weaseling out of a difficult situation; however, often such responses are actually more accurate and appropriate than conforming to the simplistic *yes* or *no* option.

Similarly, a boss will sometimes try to create a *team atmosphere* with statements such as, ***either*** *you are a team player* ***or*** *you are not a team player,* or ***either*** *you are part of the problem* ***or*** *part of the solution.* This simplistic thinking often backfires, and feelings of even greater divisiveness develop.

Obviously there are situations in which there are, in fact, only two options. Forcing someone to select one of two responses in these situations does not commit this fallacy. For example, an individual is either pregnant or not. One cannot be in an intermediate position of *being a little bit pregnant.*

As implied earlier, this fallacy is also known as *black and white thinking.* Often such black and white thinking sets up artificial situations in which only one of two extreme positions is allowed. Often that extremism is a sign of ignorance. That is, the individual who selects one extreme position is ignorant or uninformed of the many other positions that can exist.

Fallacy

of

False Dilemma

As Governor, you have a no-win situation here. Because the budget has to be balanced, either you must raise taxes or cut services. Regardless of your decision, you are going to be criticized.

The *fallacy of false dilemma* is a flawed way of thinking, in which one claims that only two undesirable options exist, when at least one additional, more-desirable option also exists.

This fallacy is very similar to the *either-or fallacy*. The difference between the two is in the type of options offered. In the either-or situation, both a desirable and an undesirable position are offered. For example, one might say *either you are my friend or you are my enemy*. Obviously selecting the friend option is not an undesirable position. In contrast, this false dilemma fallacy forces you to choose between two undesirable options—raise taxes or cut services. In both fallacies, however, more than two options actually exist.

In the example above, a third option might involve a temporary borrowing of dollars to avoid increasing taxes while also avoiding cutting services. In other words, there are more than two options, and at least one option is not necessarily an extremely undesirable position. If the economy improves, and sales tax revenues increase, the borrowed money can be quickly repaid, services can be maintained, and tax rates are not increased.

As with the either-or fallacy, sometimes there are only two options available, and sometimes both of these choices are undesirable. When this occurs, the fallacy is not committed.

Damned if you do and damned if you don't is one phrase that describes how many people feel when they perceive that they must chose between two undesirable options. *Between a rock and a hard place* is another such phrase. In psychology such a situation is simply referred to as a *lose-lose* situation.

As previously implied, a fast food restaurant can increase profits simply by assuming that customers are ordering a *medium-size meal* when clearly the advertised picture and price are for a *small-size meal.* A different fast food restaurant relies on a similar assumption, but it does so in a way that creates a *false dilemma* for its customers. On dozens of occasions after I had ordered a meal at this particular restaurant, the person at the counter has asked me, *Will that be a medium or a large?* This marketing technique is borderline deceptive because only two options are offered, although another option, *small,* is also available. In fact, just like the other fast food restaurant, this one displays both the picture and price of the *small combo meal.* Because I typically want only a light dinner, the two additional options that are offered, *medium* or *large,* are undesirable. The third option, a desirable one, is never even offered, though it is the one pictured and priced on the sign. As such, the *fallacy of the false dilemma* is committed.

When problems are examined superficially, often only two options seem to exist. There is a story about a couple who seemed to be in such a situation. This couple was expecting the birth of a baby girl, and they were discussing names for the baby. The father was insistent on *Emma,* but the mother liked the sound of *Emily.* The dilemma appeared to be that it was impossible to please both the father and the mother. Ultimately, however, the couple chose a name that was desirable for both parents. They agreed on the name *Emma Lee.* The father won because he calls his daughter by her first name, *Emma.* But, the mother, who liked the sound of *Emily* also won because she calls her daughter by her first *and* middle names, *Emma Lee.* Although such creative solutions cannot always transform apparent dilemmas into win-win outcomes, sometimes such possibilities do exist.

Often we commit the *fallacy of a false dilemma* simply because we cannot see other possibilities. When this happens, having another person look at the situation from a different perspective often produces new possibilities.

Fallacy

of

Middle Ground

My doctor gave me orders to take three antibiotic pills each day for two weeks. I honestly think one pill a day is enough. I'll compromise with him and take two pills. Everybody wins that way.

The *fallacy of the middle ground* is a faulty way of thinking, in which one automatically assumes that the midpoint between two opposing positions is the most logical choice.

Many cultures preach that moderation is a virtue. Although there is certainly some value in believing that a compromise is a good position, there are many times when a compromise is really the worst possible position. Even to say that *everything* should be done in moderation is not really a moderate position, but an extreme one.

In the example above, a decision to take just two pills per day does not simply reduce the symptoms by 50%. In fact, doing this may cause more harm than not taking any antibiotics at all. Doctors insist that we take the full course of antibiotics because a lesser or weaker dose is likely to kill off only some of the bacteria. As a result, the remaining bacteria or germs often resurface quickly and more aggressively because they have built up a resistance to the initial inadequate dose. If this happens, an antibiotic even more potent than the initial dose may be needed to stop the bacteria from growing back stronger than before.

Perhaps the most dubious use of compromise occurs when individuals purchase a new vehicle. The car dealership sets the extreme upper limit very clearly on the sticker of the car. This sticker price, or suggested retail value, is often set artificially high so that the dealer has plenty of room to compromise, but it also serves another purpose. A high upper limit often influences even the most aggressive buyers to make an initial *low* offer that is far too high. That is, if the sticker price is $30,000, many buyers think they are being very bold in offering 20% below that amount, or $24,000. In fact, many dealers act insulted or at least mildly annoyed at such an offer.

Despite such antics, a dealer may secretly be happy at your initial offer. Because the unspoken rules of the game suggest that the middle ground is reasonable for both parties, agreeing to a final price in the middle at $27,000 seems like a win-win. In fact, a buyer can boast that he forced the dealer to cut off $3,000. In reality, this was a fantastic deal for the dealer and a terrible deal for the buyer.

Most advertisements persuade us with claims about how much we can *save.* Such claims often sound fantastic, simply because the car dealer can artificially inflate the upper sticker price in a wide variety of ways. The buyer should not be impressed with how much was saved. A buyer should instead focus on how much was paid *over what the dealer paid.* (Note: The invoice price is *not* what the dealer typically paid for the car). Learning the true dealer cost, the *wholesale* price, is essential, and it can be obtained from various consumer magazines and car buyer services. More specifically, Consumerreports.com provides that information for a nominal fee.[20]

For example, my wife and I recently purchased a new vehicle that had a sticker price of over $29,000. If we had started with an initial offer of 20% off, or about $23,000, the dealer would probably have countered with an offer of about 10% off, or $26,000. Ultimately, a middle point of 15% off or $24,500 would have been reached. While this sounds like a reasonable compromise, it would have been based on an artificially inflated sticker price.

In our case, we purchased the vehicle for over $6,000 below this so-called reasonable compromise price. Ultimately we paid $18,500 for a new vehicle that had a sticker price of over $29,000. Perhaps our case was not typical; however, consumers can radically reduce the purchase price by knowing what the dealer paid for the vehicle. Based on information provided by a popular consumer Web site, and confirmed by many other Web sites, we paid about $400 over the wholesale price. Of course, we knew all of these numbers long before even stepping onto the lot. In other words, the dealer made a profit of $400. Psychologically, we can take great comfort in claiming that we *saved* more than $10,000; however, any critical consumer knows we simply made a wise purchase based on accurate information and valid reasoning.

So strong is this obligation of compromise that marketing techniques fully exploit it. Why, for example, are there often three different models, or levels, of a particular product? If only high-end and low-end models are available, most consumers struggle with making a choice. That is, they don't want to buy a model that is *cheap junk,* nor do they want to buy a model that is an *overpriced luxury.* Simply adding one or two relatively inexpensive features to the basic model often creates a mid-line model. In fact, to emphasize high quality, terms such as *good, better, and best* have been created. With these three terms, the fallacy of moderation can be practiced more effectively. Customers who initially desire the low-end model can often be pushed up to the more expensive mid-range model. This is an easy way to increase the profit margins for manufacturers while allowing customers to *select* a model that satisfies their thirst for high quality at a reasonable price.

This *fallacy of the middle ground* is also known as the *fallacy of moderation.* Like other fallacies of presumption, this fallacy begins with an incorrect assumption. Like the patient who *compromises* with the doctor and takes only half of the prescribed dose, one who seeks the middle ground without a good reason is standing on shaky ground.

Fallacy

of

Circular Argument

Jack: *It is easy to improve education—just hire the best teachers.*
Sally: *How do we do that?*

Jack: *It is easy. Just pick the best teachers to hire the best teachers.*
Sally: *But, how do we identify the best teachers so that the best teachers can hire the best teachers?*

Jack: *Like I said, the best teachers know who the best teachers are.*

The *fallacy of the circular argument* is a flawed way of thinking, in which a person attempts to prove a conclusion by incorrectly assuming the reasons used to support the conclusion have already been proven. In its most blatant form, one uses the conclusion itself in an argument to support the very conclusion one is attempting to prove.

In the example just given, Jack is simply arguing in a circle. He wants the *best* teachers to select who the *best* teachers are. Without an independent way of first determining who the best teachers are, no logical conclusion can be drawn. That is, the line of reasoning cannot continue because no foundation has been established. Without a foundation, everything built up above has no merit.

Some circular arguments are very obvious. A *history teacher* is defined as a *teacher who teaches history.* I like the song *because it is good.* Why is it good? It is a good song *because I like it.*

Although there are subtle distinctions between a *circular argument* and another fallacy called *begging the question,* most people use these terms interchangeably. As commonly understood, to *beg the question* means to *raise the question.*

For example, suppose a jury just sits down after hearing a trial and one juror instantly says, *So, should we give him the death penalty?* This statement instantly begs the question, *Is this person even guilty?* That is, one must first ask whether the question of guilt has been determined.

Although all presidents, Democrats and Republicans alike, occasionally provide circular arguments, perhaps the most obvious came from President George W. Bush on June 17, 2004:

The reason I keep insisting that there was a relationship between Iraq and Saddam and al-Qaeda [is] because there was a relationship between Iraq and al-Qaeda.[21]

Although this argument can sound persuasive when spoken with passion, it is a definitive example of the circular argument. In other words, the argument is saying that *there is a relationship because there is a relationship.*

Without any reasons or independent proof of an assertion, that assertion is simply a statement. Psychologically though, even in the absence of evidence, people are more likely to believe a statement when they hear it frequently. In fact, the fallacy *of ad nauseam,* the simple repetition of ideas as a tactic of persuasion, exploits this weakness in human reasoning.

Flawed reasoning is common in the discussions about weapons of mass destruction in Iraq. A typical question is, *So, where do you think Iraq is hiding their weapons of mass destruction?* This begs the question, *Does Iraq even have weapons of mass destruction?* Although such circular reasoning is often obvious to some, those who habitually employ such reasoning are frequently puzzled when others disagree with them.

Circular arguments are also extremely common when the topic turns to religion. There are many justifiable arguments for the existence of God; however, many people rely on faulty arguments that are circular in nature. The wording varies, however most people are familiar with the following circular arguments:

Speaker: *I know God exists.*

Listener: *How do you know that?*

Speaker: *I know God exists because it says so in the Bible.*

Listener: *Who wrote the Bible?*

Speaker: *Well, people who believed in God wrote the Bible.*

Most people see the circularity of this argument, but some do not. The next argument is far more obvious:

Speaker: *All beings were created by God.*

Listener: *Who created God?*

Speaker: *God.*

Listener: *Well, who created God's God?*

Speaker: *God.*

The following example is perhaps one of the most common examples found in logic books. Although it is obvious the following assertions do nothing more than argue in a circle, speakers in such situations frequently do not realize what they are saying.

Employer: *I can hire you if I can get objective independent proof of your honesty.*

Applicant: *Oh, you can ask my best friend here.*

Employer: *If he is your friend, how can I trust that he will be objective?*

Applicant: *Well, you can trust him. I'll testify that he is honest.*

Those who are guilty of this fallacy are often so certain in their convictions that they fail to see the emptiness of their so-called argument. In fact, they are often amazed that others could even question what they perceive as the obvious truth. In this example, the applicant doesn't realize that there is no *independent* evidence for his honesty.

Another common example is heard in discussions about crime and the law. Again, the flaw is obvious to most people; however, the one defending the position is often oblivious to the error.

Joe: *I know he is guilty.*

Cindy: *Why do you say that?*

Joe: *Well, he has got to be guilty. After all, they arrested him, didn't they?*

Cindy: *But, being arrested does not mean he is guilty.*

Joe: *I think it does.*

Cindy: *If that were true, we wouldn't need any courts or juries. We would simply convict those who are arrested because those who are arrested are guilty.*

Joe: *Exactly.*

A special type of *begging the question* or *circular argument* is known as the *fallacy of the complex question* or *loaded question.* It asks one to provide a single answer to a question; however, the question actually is more than one question. Consider the following:

1. *Have you told your parents that you flunked out of school?*

2. *When you get married, will you have children?*

In the first example, there are really two questions:

1. *Did you flunk out of school?*

2. *Did you tell your parents you flunked out of school?*

In the second example, the two questions are:

1. *Will you get married?*

2. *Will you have children?*

The reason that a *complex question* is also a form of *begging the question* is that it makes an assumption that something is true without evidence. In the first case, the question assumes that the student, in fact, flunked out of school. In the second case, it assumes that she will get married.

This fallacy is often easy to spot, because the circularity of the reasoning in some conversations is so obvious and blatant. In many cases, however, there are several steps between the initial premise and the final conclusion, so that people often don't realize that the entire argument is built on a house of cards.

This fallacy falls into the *presumption* chapter because one is assuming that the foundation on which the argument sits has already been established as being true. As is true with a house of cards, or a skyscraper that is a mile high, just one major fault in its foundation can collapse the entire structure.

Chapter 8

Fallacies of Relevance

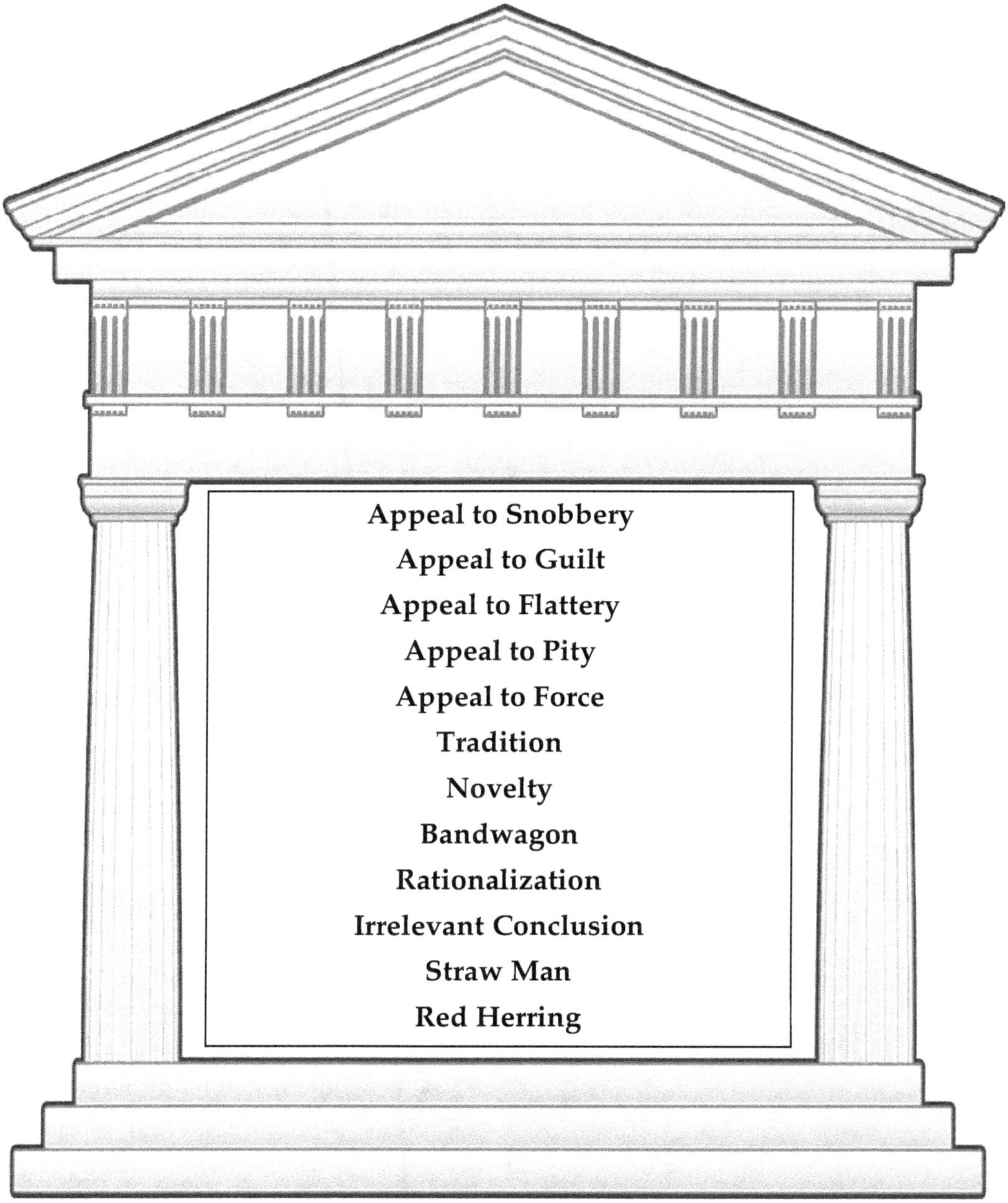

Chapter 8

Fallacies of Relevance

DRUGS, SEX, and ROCK 'N ROLL! This chant, popularized by the younger generation in the 1960s, is rarely heard today, but the force behind it is just as potent as it was decades ago. What motivates individuals to take drugs, seek sex, or listen to intense music? At the most basic level, these experiences cause emotions or feelings to change. Sex, for example, may lead to excitement, pleasure, or intimacy, or a host of other positive or negative feelings. Drugs and alcohol can also produce intense feelings of tranquility and euphoria, or of sadness and withdrawal. Music may spark pride and happiness or promote rage and frustration.

Although we rarely think about it, our desire to engage in these behaviors is an attempt to change how we presently feel. The motivation to control our feelings is a driving force behind much of what we do. Persuaders know, for example, that associating a positive feeling with the purchase of some product is often more powerful than any rational and logical list of the product's advantages. Similarly, persuaders are often successful if they can demonstrate how purchasing a product or accepting some claim erases feelings of guilt or revenge.

This chapter looks at fallacies that are related to relevance. More specifically, it looks at how people commonly use irrelevant information or feelings in making a decision. Although not all the fallacies in this chapter are related to emotions, many of them are. As such, a brief analysis of how emotions work is needed. However, before doing this, we must dispel a number of myths concerning the relationship between emotion and logic.

First, as is mentioned a number of times throughout this book, emotions are not always irrelevant to decision-making. In many social or interpersonal situations, feelings and emotions should often be considered. Such feelings as *attraction or love* are certainly relevant in deciding whether you want to marry a particular person. Informal logical fallacies occur only when emotions are used inappropriately in the process. Emotions can often pollute our reasoning, however. Substituting emotions for logical reasoning or justified evidence is a common problem. Typically, however, we are often unaware that emotions are clouding our judgments.

Second, emotions and intuition are typically processed in one part of the brain, whereas evidence and reasoning occur in another. However, there are very strong links between these two centers. In fact, a fundamental principle of psychology is that the *thinking* part of the brain can be triggered to largely control the *feeling* part of the brain. That is, rationality can control impulse.

Finally, *intuition* is not simply an emotion. In fact, many of our so-called intuitive feelings are sometimes based on logic. *A woman's intuition,* for example, is often based on logical evidence and keen observation. A mother's intuition that her baby is not feeling well is often based on evidence. The fact that the baby drank slightly less than normal before she fell asleep, or the fact that her cry was slightly higher-pitched than normal, are both logical reasons for feeling that something is wrong. Similarly, so often we hear people make such statement as, *I just can't put my finger on it, but I just don't trust her,* or *I don't know why, but I just don't feel comfortable around her.* Social science research can often provide a logical explanation for these feelings. Vocal stress, erratic eye movement, or inconsistent body language cues are logical reasons for having these feelings. So the emotional and intellectual centers of the brain are married to each other. That is, their bonds are incredibly strong.

Keep these misconceptions in mind as we now turn our attention to how emotions actually work in the reasoning process. Consider the following often-told story:[1]

> *A principal of a small middle school had a problem when some of the girls began using lipstick. The girls would apply the lipstick in the bathroom, but before leaving they would press their lips to the mirror and leave dozens of little lip prints.*
>
> *Every night the custodian would clean off the lip prints, but the girls would put them back again the next day. The principal decided this practice had to end, so he explained to the girls that it was against the rules to leave marks on the mirror. He also explained that it was a chore to remove them, and finally that putting marks on the mirror was not appropriate behavior. These reasons, however, did not seem to sink in because the lip prints would reappear every day.*
>
> *The frustrated principal asked all the girls to meet him in the bathroom one day. After the girls arrived, the custodian also showed up. The principal explained that it was very difficult to remove the lip prints, and he asked the custodian to show them just how difficult it really was. So, the*

custodian took out a long-handled squeegee, dipped it in the toilet, and cleaned the mirror with it.

Since then, there have been no lip prints on the mirror.

The principal initially gave three logical reasons why the girls should not leave lip prints, but it was the emotional appeal that finally proved effective. The principal showed the girls that kissing the mirror was akin to kissing the toilet. In other words, the emotion of disgust was more persuasive and powerful.

Like the principal, persuaders must sometimes resort to using emotional appeals to achieve logical outcomes. Emotional appeals are fallacious when they are used in place of evidence. For example, a jury should not base their verdict on whether it would satisfy an emotional need for revenge. The verdict should be based on the evidence and the law, not on satisfying emotional needs.

Although threats or blackmail are often used to intimidate others to obey or accept a demand, they are not logically relevant to the truth of a statement. Although they are psychologically relevant, and are often highly effective, they are technically not logically relevant to the claim. If your boss tells you that 2 + 2 = 5, you might agree with him if he threatens to fire you if you disagree. If he tells you that you will avoid embarrassment or humiliation if you agree, chances are you will *agree*. But, *agreement* and the *truth* are two different concepts. A fallacy is committed when satisfying an emotional need is used in place of logic and evidence in determining the truth of a claim.

Allowing emotions to overpower intellectual decision-making is a major reason why *smart* people are sometimes unsuccessful. A smart individual might do a stupid thing because his high IQ is overpowered by his low EQ, or emotional quotient. Scores of studies demonstrate that *emotional intelligence* is often a better predictor of success than IQ. Emotional intelligence means having an ability to control one's own emotions, read the emotions in others, delay gratification, and bounce back from emotional trauma.

Logic, rationality, reasoning, memory, and other intellectual functions occur primarily in the frontal lobe of the brain. Traditional education focuses on developing this area. But it is the limbic center and the amigdala, a small structure in the brain that powers feelings, instinctual reactions, and emotions. It serves as the pain and pleasure center of the brain. Quite literally, the amigdala often overpowers the logic of the frontal lobe. Being able to calm the amigdala so that rationality dominates our decision-making is a key component in overall competence.

An entire category of emotional or psychological fallacies exploits the emotional centers of the brain so that people accept arguments not based on logic and evidence, but based on pain and pleasure. The frontal lobe of the brain *knows* the potential dangers of alcohol, drugs, sex, food, etc., but the emotional center often wins that tug of war. When people get emotional, they often become stupid. That is, emotions often replace reasons.

As mentioned previously, emotions are not irrelevant to decision-making; however, sometimes we let intuitive gut feelings overpower logic and evidence. In fact, some people let the ***yuck factor*** serve as their guide. The ***yuck factor*** is the belief that one's instant reaction is an excellent guide for determining what is wise, right, or true. Although it certainly makes sense to pay attention to one's ***yuck*** response after tasting spoiled milk or smelling rotten eggs, one's physical or emotional reaction is not usually a good guide in determining what's acceptable. What made us say ***yuck*** as children might now be attractive to us as adults. Foods, customs, careers, and lifestyles that we found revolting may now seem attractive. Applying reasoning to these reactions is usually wise.

Firefighters, police officers, pilots, soldiers, emergency room doctors, and members of a host of other professions often undergo specific training to defend their emotional reactions with calm reasoning. Becoming aware of how feelings are exploited via emotional fallacies is the first step to restoring rationality to our thinking.

Desmond Tutu, the internationally known bishop from South Africa, said his father always told him: *Don't raise your voice. Improve your argument.*[2] In other words, it is important to know when we should avoid letting emotions drive our decisions. This chapter on relevance looks at how emotions, and other variables, often distract us from making wise decisions.

Fallacy

of

Appeal to Snobbery

Coach of a 14-year-old girls' volleyball team:

Congratulations, girls! Because you have displayed special talents and incredible abilities during the rigorous tryouts today, you are being invited to become a member of one of our most prestigious and elite volleyball teams. This is a special honor reserved only for the very few. In recognition of your achievement, you and your parents are requested to attend a special signing celebration tonight. Having the privilege to play on such a team is a precious opportunity, and we hope your parents will agree that the $5,000 fee we charge is money well spent. We are holding this celebration in your honor, and we look forward to having you and your parents sign the contract and financial agreement, which will allow us to reserve your spot on the team.

The *fallacy of appeal to snobbery* is a flawed way of thinking, in which one uses honor, admiration, or prestige in place of logical reasoning to seduce someone into accepting a claim.

The example above is an exaggeration, although it is based loosely on a presentation that I listened to after my daughter *made* a club volleyball team. In the example, notice the use of emotional words and phrases that suggest superiority and exclusivity: *special talents, rigorous tryouts, being invited, most prestigious and elite, special honor, reserved only for the very few, in recognition of your achievement, special signing, privilege, etc.* Imbedded in this pretentious invitation is an argument or a claim—the intent of which is to persuade the parents to make a $5,000 financial commitment. Such persuasive appeals, although largely illogical, are frequently used simply because they can motivate people to take action. Such emotional appeals exploit the weaknesses of the heart in such a way that people often ignore what the claim really is and whether any evidence is provided for the claim.

On one level, the suggestion made by the coach is that being on the team is worth $5,000. What evidence is provided for this $5,000 claim? For example, will being on the team improve my daughter's skills? Will it improve her chances of making her high school team? Will it increase the probability that she will be offered a college scholarship? Will it keep my daughter physically fit? None of these potential rational and logical arguments are given. In fact, no practical reason is provided; however, a camouflaged or sneak attack is undertaken to appeal to emotions. Making the girls (and their parents) feel *special* is the entire focus of the sales pitch. To put it more bluntly, for $5,000, we will tell you that you are *special.*

When an emotion like this is activated, people often forget about asking for evidence that shows practical benefits. Instead, they are fed words that make them drunk with warm feelings of status and adulation. In this emotional state, people allow emotions to substitute for reasons. It is at this point that the fallacy occurs—replacing logic with emotions—to make people accept a claim.

People have a strong desire to feel special–that is, to be recognized as standing out from the crowd. A great deal of advertising is aimed at making individuals feel better or superior to the masses of people. Even wealthy and sophisticated individuals fall prey to such appeals to status. Advertisements for pricey automobiles and expensive homes typically imply that these products are only for VIPs. Such products are often marketed with special wording such as: *Customers who have especially fine tastes, refined requirements, exclusive needs, those who desire only the best, etc.* The appeal to snobbery makes an effort to link the acceptance of a claim, or purchase of a product with status.

Restaurants are notorious for exploiting this emotional appeal. In order to tap the emotional needs and therefore capture the financial assets of its customers, some restaurants are found only in the most exclusive parts of many cities. In the Phoenix area, for example, Scottsdale is often considered that special place. Requiring diners to make reservations far in advance and having a dress code serve to distinguish it as an elite establishment.

And after being seated, the servers suggest not merely a soda or beer, but a specially brewed beer or a rare imported wine. Even the items on the menu are named to tickle the feelings of elegance. Everything from the appetizer, or hors d'oeuvres, to the dessert is visually and verbally presented to imply high-class sophistication. Fish eggs are called *caviar*, squid is called *calamari*, and snails are called *escargot*. All of this is done to make people feel as though they are getting the special treatment. Of course, the goal is to make them feel special and generous enough to leave a large tip.

The examples above do not suggest that appealing to one's emotions by trying to make them feel special is necessarily inappropriate or illogical. People are fully aware that they are paying a doctor or masseuse to make their bodies feel better. Those who gamble are also sometimes aware that they are paying for the opportunity to have their emotions tickled with excitement and anticipation. Fallacies occur only when we are unaware that our emotions are being simulated to overpower our thoughts. Consciously inviting emotions to dance with our thoughts is acceptable. But, when emotions sneak in the backdoor of our brains to strangle our thoughts, effective decision-making is weakened.

This fallacy is closely related to fallacies of pride, loyalty, honor, and guilt. When the passion of our pride is aroused, we accept many questionable claims. We are highly persuadable in such a state. Politicians know that sparking patriotism and pride in our nation is one of the most powerful ways of getting individuals to accept claims. In a hypnotic state, mere suggestions are often carried out without protest. Sometimes we believe a claim simply because rejecting it would arouse uneasy or uncomfortable feelings of betrayal or disloyalty. Unknowingly, many people abandon reason and evidence in their quest to capture a feeling of honor. Similarly, our brains do not carefully censor evidence when saddled with the heavy burden of guilt.

Fallacy

of

Appeal to Guilt

Car Shopper:	*No, I don't want to buy the car. I looked at the monthly payments, and I cannot afford it. Thanks for your help though.*
Salesman:	*What! I just spent the last four hours working with you, and now you are just going to walk away? I had to beg the sales manager to drop the price for you. Then I pulled all kinds of strings to get you good financing. Come on, I can't believe you would do this to me.*
Car Shopper:	*Well, okay. I suppose I can afford the payment. I guess I should buy it.*

The *fallacy of appeal to guilt* is a flawed way of thinking, in which one uses remorse, blame, obligation, or fault in place of logical reasoning to coerce someone into accepting a claim.

Guilt, like its close relative, pity, has great persuasive power because it makes people feel emotionally uncomfortable. An advertiser, for example, creates a commercial or delivers a sales pitch that conjures up some sense of internal guilt. But, he then quickly offers a remedy to numb the emotional pain—invariably the painkiller is to purchase his or her product or to make some contribution.

In the preceding example, the shopper's claim that she does not want the car because she can't afford the monthly payments is logical. However, the sales person's response is illogical, although quite typical and persuasive. Although his response was logically irrelevant because it completely ignored her concern about the monthly payments, it was psychologically relevant for her because she agreed to buy the car. She changed her mind to avoid the psychological discomfort of guilt. She had a logical reason not to buy the car initially, but that same logical reason still remained after his plea. Obviously then, she changed her mind for some reason other than a logical one. His statements were illogical, but the fact that she changed her mind based on his illogical appeal was also illogical.

Some people might argue that changing one's mind for an emotional reason is not fallacious. After all, they contend, it is logical to act in ways that reduce psychological discomfort. Although this sounds appealing, such an argument fails the basic tests of logic. Although some people might be persuaded to accept a claim to avoid pain, the actual truth of a claim is not influenced by whether pain or pleasure is associated with it. A juror, for example, might anticipate that she would feel bad after voting for a guilty verdict, but this feeling should not determine how she votes. Her vote should be based on the evidence. Similarly, the *good feeling* associated with finding someone not guilty should not be a factor in how one votes.

This is not to suggest that guilt, or other emotions, should never influence decisions or guide behavior. If someone sticks a gun in your back and proceeds to give you three reasons why you should hand your purse over, it would be foolish to base your decision on whether you thought the reasons were logical or not. The gun in your back is both physically and psychologically relevant to your decision, although technically it is not logically relevant. The mugger is not using logic, as we define it, but coercion.

Trying to logically convince a small child that he shouldn't eat all of his Halloween candy at once because it will make him sick, is bad for his teeth, and will spoil his appetite, is likely to be futile. His taste buds trump all logic. Consequently the parent has to resort to force and simply take away the candy. In this case, perhaps the small child can't even comprehend *logic.*

Similarly, the mugger will not stop his attack based on your cool-headed logical arguments. His emotions, perhaps heightened by narcotics or alcohol, control his behavior. Although trying to fight back against a mugger is not suggested, sometimes force must be met with force.

The truth of the claim, in the first example, that the car shopper couldn't afford the monthly payments was true, both at the start of the conversation and at the end. An appeal such as pity, fear or guilt, may be persuasive because it is psychologically relevant for many people, but that doesn't change the fact that it violates the rules of logic.

Although employing logic in arguments is usually persuasive for those who abide by the tenets of rational reasoning and decision-making, it is not necessarily persuasive for those who lack such judgment. As mentioned previously, logic (logos) and emotion (pathos) are persuasive tools that battle for control of our brains and hearts. Effective salespeople and politicians know this. Some people react with their hearts, whereas others respond with their brains.

Salespeople, and especially those in automobile sales, harness the potential of guilt on a regular basis. Customers are often lured onto their lot with a promise of *free* hotdogs, drinks, and balloons. In reality, these so-called free gifts cost the customers a great deal both psychologically and financially. Such gifts create a sense of obligation, a feeling that something must be given back to make it *even.* The favor of the free lunch must be repaid. The psychological principle of *reciprocity* tells us that we must return the favor. If we fail to return the favor, we are often saddled with feelings of guilt or shame.

Salespeople extract their profits by creating this sense of obligation with *free gifts;* however, they practice a game of uneven exchange. That is, if you give me a hotdog and drink, and I repay you with five dollars, I no longer have a sense of obligation. Salespeople don't allow their customers to pay off their psychological debts so easily. In fact, not only are you offered a drink and a hotdog for free, some salespeople will insist that you accept the gift. Giving such gifts creates an obligation to repay. In this emotional state, customers are far more likely to give in to the pressures of the salesperson who pushes for a higher price. That hotdog and drink, in essence, often cost hundreds of dollars.

Internal guilt or shame that a potential buyer might feel for not returning a so-called *favor* should not influence the price of the car. As such, this fallacy falls into the *fallacies of relevance.*

Fallacy

of

Appeal to Flattery

> *Mrs. Smith, the reason I was late to class was that I was in a fight. Some student said you were a terrible teacher, so I had to beat him up.*
>
> *—Unknown source—*

The *fallacy of appeal to flattery* is a flawed way of thinking, in which one compliments another in the hopes that the emotional impact of the message will be persuasive enough to overshadow the weakness of the argument, reason, or excuse.

This fallacy is almost self-explanatory and extremely common. It is just one of many fallacies that attempt to persuade by appealing to one's emotions, rather than to logic or intellectual reasoning. Essentially, the hope is to generate a positive emotional reaction or a good feeling and then use that in place of a valid reason.

Although the foregoing example of getting into a fight for the reason stated is not a logical excuse, it is psychologically difficult to disagree with someone who is complimenting you. Typically we tend to like those who show a liking toward us, and a compliment is clearly one way of showing liking. Research also indicates that we tend to be much more receptive to persuasion when it originates with a person we like.

This fallacy is very obvious when there is both a direct compliment and then a clear, but illogical, request for some action or favor. Complimenting someone about his speech, and then directly asking him to write your speech is a clear case. However, asking someone who is a very good speaker to simply help you write your speech may be a logical and sincere request for assistance. This would not be a fallacy. The fallacy occurs only when either the compliment is not genuine or when there is some ulterior motive for providing the compliment in the first place.

In the case about the teacher and student, the compliment is somewhat implied. That is, the student is suggesting a compliment by showing that he is willing to fight for the teacher's reputation or honor. In many cases, those who are guilty of this fallacy will camouflage their compliments so as to avoid being too obvious or arousing suspicion.

Complimenting someone as a way of getting that individual to agree with a particular argument or excuse is psychologically very powerful. This exploits the principle of reciprocity, which basically obligates one person to another when a gift (or compliment) is given. This sense of obligation, as explained in a previous fallacy, is widely used in sales in which some token gift is given early in the process. Sometimes the gifts are tangible—a free calculator, address labels, memo pads, tote bag, makeup samples, or hundreds of other inexpensive items. Often, however, the obligatory gift comes in the form of compliments. Many salespeople are quick to compliment potential customers on everything from their dress to their skillful negotiating abilities. This flattery serves a very useful persuasive purpose.

However, regardless of how much flattery one might use, these compliments should not determine whether to accept an argument. Being showered with compliments should not influence, for example, how one determines a fair price for a new car. Always being cynical or suspicious of people who give you genuine compliments is not psychologically healthy. But, we should be somewhat skeptical of those who seem to pass out too many compliments, especially if those people are seeking something that we have and they want.

Fallacy

of

Appeal to Pity

Attorney to jury: *The defendant was abused as a child and grew up in one of the toughest ghettos in the country. His father left when he was just 12 years old, and his mother has been in and out of prison six times for selling drugs. Yes, he is an alcoholic, but he is going to rehabilitation twice a week. He is a good kid, and he is trying his best. Given these circumstances, you should find him not guilty of drunk driving.*

The *fallacy of appeal to pity* is a flawed way of thinking, in which one tries to generate sympathy so that decisions will be based on feelings, rather than on reasoning, logic, or the law.

Like the fallacy of flattery, this fallacy also attempts to make people feel a certain way so they are more likely to make decisions or take actions based upon achieving emotional comfort rather than on logical reasoning. This tactic is often highly effective because humans are not purely rational beings devoid of emotions or feelings. Many persuaders are highly skilled at getting people to listen to their hearts or emotions, rather than to their brains or intellect.

In the case above, it is perfectly natural and logical to feel some compassion for the defendant. Relying on one's emotions to make decisions in certain situations is appropriate. As mentioned previously, deciding on whether to accept a marriage proposal should involve both one's emotions and intellect.

Having sympathy, showing compassion, or feeling sorry for people is both humane and logical. It is not logical, however, to rely on these emotions to make a decision that is supposed to be based on evidence and the law. In other words, such emotions should not guide a jury in determining innocence or guilt. The evidence and logical reasoning should be used in making this determination. At this point, any sympathy generated by learning of the defendant's past is not relevant to whether a crime was committed or not. The defendant's past, along with compassion and sympathy, may be logically considered in the penalty phase of the process, however. Whether this is the first offense and whether the defendant is serious about rehabilitation are extenuating factors that are often considered.

The fallacies of pity, flattery, guilt, and several other psychological appeals all attempt to intensify feelings so that the emotions of the heart become more powerful than the reasons of one's brain. These are all categorized as fallacies of relevance because frequently emotions are not relevant to the truth of a claim.

Fallacy

of

Appeal to Force

Waitress:	*Because I have already worked 60 hours this week, and because my son has a soccer match, I won't be able to come in to work this Saturday.*
Boss:	*Okay. That sounds reasonable.*

The *fallacy of appeal to force* is a flawed way of thinking, in which one uses physical force or psychological fear in place of logical reasoning to coerce someone into accepting a claim.

Although virtually every entry in this handbook begins with an example of a fallacy, there is no such fallacy in the above conversation. In order to understand the *fallacy of appeal to force,* it is often necessary to first start with an argument that most agree is a sound or legitimate argument. Logical reasoning and legitimate reasons are used above; consequently the argument is a strong one. We will contrast this with a fallacious argument a bit later.

The fallacy itself is extremely common, and almost everyone is guilty of occasionally relying on it. If we cannot accomplish what we want through rational means, we often resort to force. This fallacy is committed in various ways. The most obvious way to get someone to accept a claim is by using sheer physical force. If a schoolyard bully, for example, doesn't get his way when he asks a timid student for his lunch money, he may resort to twisting the boy's arm until he screams. To avoid even more pain, the boy habitually relinquishes his lunch money. Often, simply the threat of being beaten up will achieve the same result. Threats of blackmail or screams of intimidation are other types of psychological force.

Let's return to the original example, but now we will make a few additions to more clearly illustrate the fallacy. Pay special attention to those words that are printed in bold.

Waitress: *Because I have already worked 60 hours this week, and because my son has a soccer match, I won't be able to come in to work this Saturday.*

Boss: *Okay. That sounds reasonable.* ***Oh, by the way, I will be completing employee reviews next week. I hope we won't have to lay off anyone.***

Waitress: *On second thought, my husband can take him to the match. I'll see you bright and early this Saturday.*

In this fallacious example, note that the reasons the employee gives for not working the weekend are still legitimate reasons. The boss makes no effort to counter the reasons provided by the employee. He does not, for example, indicate that she must work because other waitresses are out sick. Nor does he say that the busy holiday weekend will make the restaurant especially busy. Both of these would be legitimate reasons.

Instead, he tries to get her to accept the fact that she has to work this weekend by making a subtle threat that she may lose her job if she does not comply or obey. This threat is not a *logical* reason, nor is it logically relevant to the situation. The threat, of course, is psychologically relevant, but not logically relevant. In other words, using this threat violates the rules of logic, despite the fact that it might be effective in achieving what he desires. He has, in effect, broken the rules of logic. Although there are rules for how prisoners of war are to be treated based upon the rules agreed to at the Geneva Convention, many countries sometime ignore the rules. Ignoring the rules does not change the rules, just as using force, intimidation, or threats do not change the truth of an argument. People *accept* such claims to avoid physical or psychological pain, not because the claims are true.

In 1633, for example, the famous astronomer Galileo was threatened with death when he announced that he could prove that the earth revolved around the sun and not the other way around. Because various religions believed that the earth was the center of the universe, his proof was buried for years. Although Galileo was forced to recant his discovery and *accept* the Church's version of the world, the threat of death did not change the truth of his claim. The threat only forced Galileo to publicly *accept* the Church's claim.

In determining whether one should accept a claim, it is often necessary to remove any positive or negative consequences associated with either accepting or rejecting the claim. For example, in determining whether there will be another terrorist attack similar to the 9-11, one cannot logically say: *There will never be another major terrorist attack in the United States because I simply cannot bear to think of how awful that would be.* We cannot deny the truth of some claim just because it would mean unspeakable pain and suffering. Some people say they are certain that a nuclear bomb will never be detonated because the destruction, loss of life, pain and suffering would be impossible to bear. Just because we cannot handle the truth doesn't affect the truth itself or the probability that a horrendous event might occur. The truth of a claim is not influenced by whether it would cause pain or pleasure.

Some politicians have been quite successful in getting elected by creating an atmosphere of fear. And, once in a fearful state, people often accept the flimsiest of arguments—arguments without logical reasoning and without credible evidence. In other words, people no longer behave based on rationality, but instead react based on emotions and feelings. And, as suggested earlier, intense feelings are often far more predictive of behavior than is rationality. A feeling of security or safety, for example, is a powerful human need. Often the personality of a politician can make people *feel safe,* regardless of whether there is evidence to show that his or her actions or policies have increased security. This is not to suggest that gut feelings or intuition should be entirely dismissed in judging others in all cases; however, many people rely on emotions far too extensively, and even believe in the *truth* of these emotions in the face of overwhelming evidence and sound counterarguments.

Logicians do not suggest that force, or the threat of force, is always inappropriate in order to get someone to obey or comply. Using force to get others to comply with a law is perfectly logical in a wide variety of cases. What is not logical, however, is to assume that force has anything to do with the actual truth of a claim. If police detectives use torture to extract a confession from a suspect, the torture (or force) serves only to obtain obedience, compliance, or admission of guilt. Although force or pain may be effective in getting a suspect to *confess* to a crime, the actual truth is independent of whether force was used.

Sound arguments depend on logical reasoning and justified (or true) premises. If threatened or tortured, I might *accept* that the world is flat, *claim* that the world is flat, or maybe even *believe* that the world is flat. But, none of this changes the fact that the world is not flat.

This fallacy falls into the *fallacies of relevance* chapter because it seduces people into accepting a claim, not based on sound reasons and evidence, but on the fact that one will feel psychological or emotional comfort by accepting it. More specifically, this fallacy attempts to turn feelings of fear into feelings of safety or security. A frightened child can often be comforted by a reassuring word from a parent. But, adults should not accept such reassurance based on blind faith. A politician who ensures that he will keep us safe has an obligation to show us how and why his plan will achieve that goal.

Fallacy

of

Tradition

Son of Rancher:	*Why do I have to get my hair cut so short?*
Rancher to Son:	*You need to have short hair because I have short hair, and your grandfather has short hair. That's the way it's always been—boys have short hair, and girls have long hair.*

The *fallacy of tradition* is a faulty way of thinking, in which one argues that the ways of the past are best, or should be accepted without providing a rational reason why.

Ask your boss why something needs to be done in a certain way, and frequently the response is *because it has always been done that way.* Nearly all parents are guilty of using the standard line, *when I was your age,* or *when I was a kid.* Just because something has been done a certain way for years is not, in itself, justification for why it should continue to be done in that way.

This is not to suggest that the ways of the past, tradition, or older practices should be abandoned. Quite the contrary—many practices of the past should continue because they have been shown to be effective or efficient. That is, they have been time-tested. Wine that is aged, for example, usually becomes better. Milk, on the other hand, does not. If a logical rationale explains why the practice should continue, a fallacy is not committed.

In the previous example, a fallacy occurs because no reason is provided. There are many rational reasons why a rancher's son should have short hair. In addition to the fact that it is easier to care for, having short hair around animals is far healthier and safer than having long hair.

Those who say that it has been *practiced for decades* or has been *believed for hundreds of years* often don't consider the fact that many such practices or beliefs have been shown to be harmful or wrong. In medicine, *bloodletting,* the act of draining blood from those who were sick, was practiced for decades; however, modern medicine knows it is incredibly harmful.

Older job applicants might tout their *years of experience* as a rationale as to why they should be hired; however, if they don't explain what skills they developed as a result of those years, a fallacy occurs. There are millions of examples from business, education, and politics in which older or more experienced candidates failed to show competence. The number of years of experience is not nearly as significant as the wisdom or skills that were attained during that time. There are those who repeat the same mistake dozens of times and never change their ways. In contrast, there are those who have less experience, but who quickly learn from their mistakes.

In the same way, the fact that a particular technology has been around a long time does not mean that it will continue to exist. For example, Polaroid employed more than 20,000 people worldwide in 1978,[3] but the company failed to quickly embrace the idea that the traditional way of taking pictures—with photographic film—would soon be obsolete. Similarly, Kodak reduced its workforce by 10,000 employees because it failed to see that taking photographs in the traditional way would soon be replaced with digital technology.[4]

There is an old story about a woman who always began cooking a ham for dinner by first cutting off the ends of the ham. When a friend asked her why she did this, she simply replied that her mother always did it that way. When her mother was asked why she did this, she replied that her mother had done the same thing. Finally, when asked why she had cut off the ends of the ham, the great-grandmother replied that she had done so because the ham would not fit in her pan otherwise.

So, the great-grandmother didn't commit a fallacy, because she had a rational reason for doing what she did. However, all of the others were guilty simply because they blindly followed past practice and never asked why.

This fallacy also falls into the relevance chapter simply because the age of an idea is not necessarily relevant in determining whether a claim is true. This *fallacy of tradition* is the converse of the *fallacy of novelty*, the next fallacy we will address.

Fallacy

of

Novelty

Microsoft just released Vista, its new operating system, and I can't wait to get my hands on it. I know it is going to be so much better than Windows XP.

The *fallacy of novelty* is a faulty way of thinking, in which one automatically believes a claim should be accepted simply because it is new.

This is one of the simplest fallacies to understand, but it is also one of the most frequently committed. It is the opposite of the *fallacy of tradition*. In this example, whenever a new operating system is released, there are almost always widespread reports of various bugs, flaws, or disadvantages. In fact, many users of the new system often initially wonder whether it was a wise move to upgrade. Because some users are using older computers, the new system might actually be slower and less compatible with some older hardware. Security issues and various bugs are common in almost all new upgrades. Buying the new system will obviously cost the user more money, and installing it will probably take substantial time and cause some headaches.

Buyers who are the first on the block to own the new technology (new cameras, TVs, computers, cell phones, etc.) are also those who typically pay high prices for those items. After a product is on the market for a few years, prices for these new technologies often decline.

Some might argue that being the first on the block to own a new gadget brings with it the attention or admiration of friends. Although this is true, buying a product simply because it increases one's status may involve the *fallacy of flattery.*

With respect to automobiles, the first cars off the assembly line with a newly redesigned model or style are also those that experience more reliability problems. Because these flaws are usually fixed the following year, waiting a year or two after the car's introduction can often avoid these problems. Again, newness does not necessarily mean better.

Similarly, many patients who see new medications advertised in the media are often the first to pressure their doctors to write prescriptions for the new drugs. Although such novel medicines often provide relief for many patients, the fact that the drugs are new also carries with them the possibility of unknown detrimental long-term effects.

This is not to suggest that new products should be avoided. The initial period of problems is the cost we pay for advancing technology. After a period of time, and after a period of improvement, it is true that most of the new products will turn out to be *better* than the old products. One commits the fallacy only when one automatically believes that *new* instantly and automatically means better.

In some cases, a new product turns out to be so dangerous or flawed that it is removed from the market. Similarly, the sign that reads *under new management* implies that customers will receive better service than they received in the past. Sometimes this turns out to be true, but sometimes it does not.

Advertisers know that many people are easily persuaded by such words as *new, just introduced, just arrived, the latest,* or any number of similar words that suggest novelty. However, wise consumers can often resist this appeal.

Fallacy

of

Bandwagon

> *Of course I believe I have a guardian angel. You hear people talk about having an angel looking out for them all the time. Almost everyone I know thinks they have a guardian angel, . . . and an article in the paper said that two out of three people believe in guardian angels. Obviously, because so many people think they exist, they must be real.*

The *fallacy of the bandwagon* is a faulty way of thinking, in which one believes that a particular claim is true or should be accepted *simply* because is it believed by a large number of people.

The masses of people once believed that the world was flat. They also believed that it was at the center of the universe. Both statements have proven to be incorrect.

Typically, we admire those people who are riding the floats in a parade. Musicians or band members, for example, often ride a wagon in the parade while performing. The *bandwagon fallacy* is derived from the idea that those on the bandwagon are popular and are to be admired. That is, we all want to be like these popular individuals who are riding the bandwagon, and thus we often believe or act as they do.

Although it is generally a fallacy to accept a claim simply because it is popularly believed, there are times when popularity may be relevant to validity. For example, the Honda Accord is a very popular vehicle; however, one must admit that the reason for much of this popularity is due to its high reliability. You would be committing a fallacy if you bought a Honda *solely* because it is so popular, but if you bought it for the reason of reliability, you would not be committing a fallacy. So, just because you happen to buy a popular product or accept a popular claim does not necessarily mean that you committed a fallacy.

In some cases, popularity *is* a relevant reason for accepting a claim. For example, if the masses of people truly believe an economic recession is impending, their popular belief will influence their behavior in such a way that they will create a recession. Believing a recession is near is likely to cause people to buy fewer products or to not invest in new equipment. This slowdown in buying, then, is likely to cause a recession.

Similarly, the popularity of a president or manager, for example, influences to some extent, whether he or she can truly be effective. If a president does not have the confidence or support of the masses of people, the ability to be effective is greatly hampered. Popularity is sometimes one criterion for determining effectiveness. Many entertainers, for examples, are considered effective simply because of their popularity.

Despite these examples, popularity, in and of itself, is usually not sufficient reason for accepting a claim as being true or valid. Holding a correct view and holding a popular view are often at odds.

This fallacy falls into the *fallacies of relevance* simply because popularity is often irrelevant to the truth. Though not logically relevant, sometimes this argument is persuasive because many people allow it to be psychologically relevant. That is, one psychological need is to gain the approval of our friends, and one way to gain their liking is to do or believe what they do. In many ways, this is a type of peer pressure. When our peers are doing something, we too feel a need to follow along. Simply the desire to be in the majority, or on the winning side, creates so much emotional force that people often disregard logic and evidence when making a decision.

Fallacy

of

Rationalization

Teenage Girlfriend:	*Did your parents give you a gift for high school graduation?*
Teenage Boyfriend:	*Oh yeah! I convinced them to buy me a Hummer SUV. After all, it is such a practical vehicle. It is big and safe. It is also made in the U.S.A.*

The *fallacy of rationalization* is a flawed way of thinking, in which one presents logical or acceptable reasons for an argument, while concealing the underlying or genuine motivating reasons for the argument.

In the example above, the argument that this vehicle is big and safe is probably one that his parents might find logical. Even the fact that it is made in America might make the argument more acceptable. We cannot say with certainty what the underlying or genuine reason is as to why this high school graduate wants a Hummer; however, it is doubtful that his expressed reasons are the real reasons. In all probability, the overwhelming reason that an 18-year old desires this vehicle is because it is *cool.* It clearly attracts attention and is a status symbol. Although these are the genuine reasons, obviously these emotional reasons are never expressed because it is unlikely one's parents would pay for such an expensive vehicle for these frivolous reasons.

Even casual conversations are often interspersed with this fallacy. The coed who is asked for a date responds with a persuasive, yet untrue argument: *Sorry, I have a boyfriend.* Wanting to take a day off to play golf, a worker conjures up an argument that sounds rational and acceptable: *I feel really sick today, so I won't be coming in to work.*

Although we often use rationalizations consciously, we sometimes use them unconsciously as a type of self-deception. Suppose in the example above, the young man proposes to his girlfriend. Assume further that he has been convicted of selling drugs, had spent more than six months in jail, has fathered two children with two different women, and has never kept a job for more than two weeks. Despite this, she eagerly accepts his proposal. She *loves him,* and she clings to the rationalization that *love conquers all.*

Such rationalizations are common in relationships; however advertisers and politicians rely on them as well. These persuaders know that emotional appeals are often more powerful than appeals to intellect. Many consumers buy on impulse and emotion. Later, when explaining the purchase to a spouse or friend, buyers tend to stress the logical reasons for the purchase. That is, people generally want to be perceived by others as logical and rational.

In politics, the underlying reasons for various military actions are often called into question. The masses of people in the United States probably could not be easily convinced that its military should invade another country to access its oil. In contrast, if the invasion is based on the stated desire to stop the spread of weapons or to prevent terrorism, the public might be more receptive. Whether the official reason is the real reason, or just a rationalization, is hard to know with certainty.

This chapter has focused on how irrelevant elements are used to make claims. The *fallacy of rationalization* sometimes hides the genuine emotional forces behind our decisions. A number of political examples in this book have illustrated how emotions and rationalization intersect. Simply bringing up a political issue often incites such intense feelings in some people that having a rational discussion is next to impossible.

Despite this, ignoring one of the most important political and economic issues of our generation, the war in Iraq, would be inexcusable. Discussions about the war have provided a wealth of information about how people reason, both correctly and incorrectly. What people say or write about the war can be analyzed for errors in reasoning; however, it is often what is *not* said that drives actual decisions.

The stated goal for the war was to destroy the weapons of mass destruction. Because there were no such weapons, the goal became a moot point. But, is it possible that this *fallacy of rationalization* covered up other equally important, but unstated goals? The war had the potential for huge benefits, but it also had a high risk of catastrophic failure. Was one unstated goal to establish a stable democracy in the Middle East? Was there a desire to send a message to the world that the United States would respond aggressively to perceived threats? Was there an unspoken goal to establish a relationship with a country that could potentially provide a vast source of cheap oil? Although all of these are legitimate reasons, they were not the reasons used to sell the war to Congress and the American public.

Popular opinion and political pressures obviously influence decisions. The White House administration lobbied Congress to pass a resolution authorizing the President to use military force in Iraq. The majority of his party, and several from the Democratic Party, voted for the resolution. Did they pass this resolution for logical reasons? Did they pass it for political reasons? Did they pass it out of loyalty? Does loyalty actually mean obedience? Does loyalty obligate one to question evidence or reasoning that is weak?

It is impossible to provide any definitive answers to these questions. Similarly, it is impossible to identify all the underlying and unstated reasons for the war. It is also impossible to know exactly why so many in Congress did not scrutinize the CIA's flawed conclusion that Iraq had weapons of mass destruction. Was it a failure of critical thinking?

In the case of the war with Iraq, evidence indicates that some decision-makers failed to critically analyze both the evidence and the data. To repeat, the report produced by the Senate in 2004 was very clear.[5]

> *In a hard-hitting report released today, the U.S. Senate's Select Committee on Intelligence said the CIA and other agencies used unfounded "group think" assumptions to assess the existence of weapons of mass destruction in Iraq before last year's U.S. invasion and reached conclusions that were often either "overstated" or "not supported by the underlying intelligence."*

Criticizing a decision *simply* because it led to disastrous results is not necessarily the sign of a critical thinker. There are many decisions that are logical and rational that produce terrible results. One cannot look at the results alone and state unequivocally that flawed reasoning was to blame. People can bet on the favored horse, for example, and still lose. On the other hand, a positive outcome is not necessarily the sign of sound thinking. A gambler can unknowingly make a bet against great probability and still win just by *dumb luck*. Avoiding the fallacies in this book will not guarantee great results. Critical thinking simply *improves the probability* of a successful outcome.

Many opposed the war long before it started because, as they pointed out, the evidence was very weak, and the reasoning was questionable. They argued that the decision-makers favoring the war were committing groupthink, apriorism, and several other errors. Because they saw the flawed reasoning as it was unfolding, they were not especially surprised by the poor outcome. They concluded that the probability for success was not worth the potential costs and risks.

In contrast, some individuals began to oppose the war only after it became clear that the results were not promising. Changing one's mind because of a poor result is not an example of pristine logic. In contrast, changing one's mind after realizing that the decision-making *process* itself was flawed is far more acceptable. In legal cases, for example, a losing attorney files an appeal, *not* because of the verdict itself, but because he or she suspects that a *faulty process* was used in making that decision.[6]

Although there is no way to completely defend against those who employ these errors of rationalization, one can investigate whether the evidence really supports the stated argument. Being aware that many people are strongly motivated by political, religious, or personal interest is often the first step toward fighting this fallacious form of reasoning.

Fallacy

of

Irrelevant Conclusion

> Arizona Politician: *When compared with all the other states on total dollars spent on K-12 education, we usually come in about average, normally about 21st. Therefore, I think it pretty clear that we have adequately funded education in this state.*

The *fallacy of the irrelevant conclusion* is a faulty way of thinking, in which correct and accurate information is used in the premises of an argument; however, the conclusion does not follow logically from those premises. Sometimes this is called *missing the point*, or a *non sequitur*, which means basically *it does not follow.*

Put another way, one is trying to make a claim and carefully selects information that can be verified, but the reasoning applied to these premises does not rationally lead to the stated conclusion. In fact, the premises lead to a different or irrelevant conclusion.

In the previous example, the fallacy of the irrelevant conclusion occurs when a politician suggests that *K-12 education spending in Arizona* is about *average* because it is *ranked 21st out of 50 states on total dollars spent.* This is an inaccurate and irrelevant conclusion, because Arizona has over *a million* more people than does the state that is ranked 21st in population.[7] Consequently, if one looks at the *total dollars spent,* Arizona is about 21st.[8] However, when ranked *relative to population (that is, relative to dollars spent per student),* Arizona drops from 21st to 49th or 50th.[9] Using total dollars without factoring in population produces conclusions that are irrelevant and inaccurate.

Many suggest that this simple ranking is not a fair picture of educational achievement in the state of Arizona. In fact, by some measures, Arizona's schools are highly ranked. Some scales indicate that Arizona students do better than the national average. *This is not the point.* The point is that *one cannot logically conclude* that educational funding is *about average* because a state *ranks 21st* on total dollars spent. Such a ranking would be terrible for California, the most populous state; however, the same ranking for Wyoming, the least populous state, would be exceptional.

Statistics concerning education have to be carefully analyzed to produce legitimate conclusions. Similarly, crime statistics are often used to produce dubious conclusions. For example, consider the following information:[10]

Murders in the United States in 1970: 16,000.

Murders in the United States in 2007: 16,929.

Although these statistics are correct, it does not logically follow, as some suggest, that the murder rate has gone up during this time. In fact, consider that in 1970 that there were nearly 100 million fewer people in the United States than there were in 2007. Although the total number of murders went up by about 10%, the population increased by 50%. As such, the murder rate did *not* go up—it went *down dramatically*.

Political parties also use rankings or statistics in inappropriate ways to convince voters to accept conclusions that are highly dubious. That is, true information is used, but the reasoning applied to the information does not lead to the claimed conclusion. A rather insidious political claim circulating on the Internet is one that *FactCheck.org* and *www.snopes.com* expose as a fraud.

The basic claim is that in the 2008 presidential election, the square miles of land won by Democrats was about 600,000. The square miles won by Republicans was about nearly four times greater, at 2,400,000. Based on these facts, some have asserted that the republican nominee, rather than the democratic nominee, should be the rightfully elected president. This is an irrelevant conclusion simply because votes are based primarily on population (electoral votes), not on square miles. In fact, if ownership of land by voters were used, CNN founder, Ted Turner, who is the largest private landowner in the country, would have greater voting power than the *entire population of the 10 largest cities in the country.*[11] Turner owns nearly two million acres.[12]

Such irrelevant conclusions often emerge from those who have ultraconservative or ultraliberal political ideologies. Extremism, whether it is from the right or left, often seduces people to accept conclusions that moderate voters would analyze with more skepticism.

Fallacy

of

Straw Man

Jack: *I think the war in Iraq was a mistake.*

Sally: *I can't believe you are so against our troops. I don't know why you dislike our young men and women who risk their lives everyday.*

The *fallacy of the straw man* is a faulty way of thinking, in which one attempts to gain some advantage by misrepresenting an opponent's position, or by refuting the weakest part of the opponent's argument and claiming total victory.

This fallacy, as the name implies, is one that relies on appearance rather than on the facts. Like a scarecrow, the straw man is a misrepresentation of an argument in such a way that the argument appears weak. A weak argument, like a straw man, is easy to knock down. In the classic movie, *The Wizard of Oz,* the scarecrow was a man without a brain. In a sense, one who set up a straw man paints his opponent as holding a mindless or ridiculous position.

Like the birds that are often scared away from a fake man made of straw, individuals are often frightened away from arguments that are misrepresented or unappealing. For example, nearly everyone acknowledges the huge personal sacrifices the troops are making in the war in Iraq. That is, virtually no one is against the troops. However, many who are opposed to the war have been mischaracterized as being unpatriotic and against the troops. Unfortunately many people react instinctively to a negative or repulsive image. That is, they don't take the time to determine whether the image or position is being presented accurately.

As the straw man metaphor implies, a strong man, like a real argument, is far more difficult to defeat than is an imitation of the real thing. In fact, because it is nearly impossible to defeat some strong arguments, those who wish to knock down these arguments often resort to devious, but highly effective, ways of distorting the arguments. Rather than legitimately trying to defeat a worthy opponent, some people will either blatantly misrepresent or exaggerate an opponent's position so that it is easier to defeat.

Many politicians often rely on using the straw man. For example, a candidate may endorse a pro-choice position; however, he or she is often misrepresented as being a strong supporter of abortion as a routine method of birth control. Just as comedians have dozens of jokes ready to entertain, many politicians also have dozens of straw man attacks ready to unleash on their opponents. One reason these attacks are so successful is that many people hear and see nothing more than a headline in the newspaper, the sound bite on the radio, or a blurb on a Web site.

Rarely do people take the time to do their own research to determine whether these statements are accurate or misrepresentations. Negative advertising is notorious for taking a candidate's position out of context or exaggerating it in such a way that most people find it repulsive.

Frequently those guilty of using a straw man are fully aware that their charges are misrepresentations. They also fully realize that some people may contest their falsified claims; however they know that most will not. Regardless, they know they have inflicted damage. Even when compelled to apologize or retract their charges, they know that they have planted a seed of doubt in the public's mind.

Consider the following examples:

Jack: *I am opposed to a law banning the burning of the American Flag.*

Sally: *I can't believe you hate our country so much that you want everyone to burn the flag.*

Tom: *I support laws that allow all people, regardless of sexual orientation, to marry.*

Carol: *I can't believe you are against a man and woman getting married and having children.*

Joe: *I think it is essential that kids in high school be taught about birth control.*

Mary: *I can't believe you think it is a good idea for high school students to have sex.*

One version of the straw man fallacy occurs when one simply defeats, or knocks down, only a small or insignificant part of an opposing argument and then claims that the entire argument was defeated. For example, consider the following rather extreme example:

Prosecution: *Ladies and gentlemen of the jury, we have shown that the defendant is guilty beyond a reasonable doubt. Overwhelming physical and eyewitness testimony proves he committed the murder late on the night of December 31.*

Defense: *Ladies and gentlemen of the jury. The defendant is not guilty. We can prove that he did not leave his home until midnight on the night of the murder.*

This is an extreme example of setting up a straw man by quibbling about a minor detail. Technically, one part of the accusation was refuted, the precise minute the murder occurred. But, refuting such a minor part of an argument is not grounds for suggesting that the entire claim can be dismissed.

In most cases, the straw man fallacy occurs in more subtle ways. Typically several arguments are made for a case, some strong and some weak; however, being able to refute the weakest of all those arguments is not grounds for claiming that the entire charge should be dismissed.

This fallacy falls into this chapter on relevance simply because the misrepresentation of the straw man is not pertinent or significant to the truth of a claim. Of all the fallacies discussed, this is one of the most popular because it is easy to commit and often difficult to detect.

Fallacy

of

Red Herring

Reporter:	*Was the war in Iraq a mistake?*
Politician:	*I don't think it was a mistake to remove Saddam Hussein. He was a brutal dictator that killed thousands of his own people.*

The *fallacy of the red herring* is a faulty way of thinking, in which an irrelevant answer or statement distracts people from remaining focused on the original issue.

In the example above, the supporter of the war dodged the question by changing the subject. Rather than answering the question of whether the war was a mistake, he instead answered the question, *was removing a brutal dictator a mistake?* These are two entirely different, although seemingly related, questions. When people wish to shift the topic of conversation or avoid answering an unpleasant question, they often resort to using red herrings. Rather than having to admit a mistake or support an unpopular position, the red herring response allows people to deflect the question, answer a different question, or side-track the discussion.

A politician at a new conference, for example, will often spend several minutes on a red herring diversion. The goal is to speak long enough so that people will lose track of what the original issue was. For example, in the example above, it would be easy for virtually any politician to talk at length about the evils of a brutal dictator. If the red herring digression takes substantial time, the reporter often is not allowed additional time for a follow-up question or to suggest that the original question was never answered.

Like poorly trained bloodhounds that are easily led off the trail, the general public has a short attention span. The term *red herring* comes from the practice of using dogs to track the trail of an escaped prisoner. Bloodhounds can follow the scent of an escapee for miles if they are well trained. To test the dog's ability, a trainer would pull a red herring (a smelly rotten fish) across the original trail to see whether the dog would stay focused on the original scent or be distracted by the new odor.

Notice how the speaker below attempts to divert the attention of the listeners:

Reporter: *Are you going to cut funding for education?*

Politician: *Education is a priority, and everything will be done to maintain that high educational standards are the norm in this state. Although our tax revenues are low, we must recognize how dedicated our teachers are in teaching our students with limited resources . . .*

Because all politicians must occasionally take positions that are unpopular with some people, they often rely, consciously or unconsciously, on the red herring to sidetrack the issue. As in this case, the question goes unanswered; however a lengthy monologue about teachers, education, and students ensues. The red herring is sometimes challenging to detect because listeners cannot instantly determine whether they are being led astray or not.

Red herrings can also simply be witty remarks, jokes, or subtle insults. To avoid being drawn into discussing the merits of the pro-choice stance, Ronald Reagan, 40th President of the United States, remarked:

I've noticed that everybody that is for abortion has already been born.[13]

This fallacy falls into the relevance chapter because one who commits it is providing a response or answer that is not logically pertinent to the original issue. Sometimes this is called simply *a wild goose chase, smoke and mirrors,* or *clouding the issue.*

Chapter 9

Final Chapter

Cognitive and Perceptual Mistakes

Depressive Realism

Positive Illusions

Final Thought

Chapter 9

Final Chapter

A mind once stretched by a new idea can never go back to its original dimensions.[1]

—Oliver Wendell Homes (1809–1894)—

If this statement is true, perhaps a few of the concepts in this book have already irrevocably altered some minds. In other words, the damage has been done, and there is no way to repair it or reverse it. Damage inflicted against an enemy is, of course, desirable. Destroying infectious reasoning illusions and viral thinking delusions is precisely the intended goal. A regular dose of critical thinking can help us achieve this goal, but can this medicine produce toxic side effects? That is, can it produce feelings of chronic cynicism? Do the side effects outweigh the benefits?

Cognitive and Perceptual Errors

One benefit of critical thinking is the sharpening of our intellectual skills to make better decisions. However, critical thinking can also improve our psychological understanding of communication and human behavior. But, critical thinking also demands that we also consider potential negative side effects of such thinking. Before we consider the negatives, however, a review of almost any college textbook in social psychology reveals that most people routinely commit scores of cognitive and perceptual mistakes. Such research indicates that most people:[2]

1. Believe that they have more control over events than they actually do.
   (Illusion of Control Effect)

2. Believe with extreme confidence that their answers to many questions are correct (when they are not).
   (Overconfidence Effect)

3. Believe others agree with them more than is actually the case.
   (False Consensus Effect)

4. Believe that the probability that good things will happen to them is higher than it actually is. They also underestimate the probability that bad things will happen.
   (Positive Outcome Bias)

5. Believe that they have higher competence than they actually do. They also fail to have the competence to realize their incompetence.
   (Dunning-Kruger Effect)

6. Believe themselves to be better than the average individual on a wide variety of characteristics.
   (Illusory Superiority Effect)

7. Believe that they are more responsible for desirable outcomes than they actually are. They also blame others more for undesirable outcomes than is the case. That is, they take the credit for success, but avoid blame for failure.
   (Self-Serving Bias)

A simple Internet search of *cognitive and perceptual errors* will reveal several dozen such mistakes, along with supporting studies. For example, there are dozens of studies like the one below that serve to document the *Illusory Superiority Effect.*

In one study using a random sample of men, the question was posed, *how would you rank your ability to get along with others?* Of the sample, 60% ranked themselves in the top 10%, and, amazingly, ***a full 25% ranked themselves in the top 1%.*** When asked about athletic abilities, only 6% ranked themselves below average. In one study of nearly a million high school seniors, 70% said they had above average leadership skills, but only 2% felt their leadership skills were below average.[3]

In looking at our inflated opinion of ourselves, perhaps Father Larry Lorenzoni said it best: *The average person thinks he isn't.*[4]

Perhaps it is not surprising to find evidence indicating that many people have inaccurate thoughts and perceptions. However, one should be surprised at the extent to which his or her perceptions depart ***radically*** from reality.

So, given the fact that such errors are common, a dose of critical thinking is called for. However, critical thinkers also carry a burden of side effects.

Depressive Realism

The quote at the beginning of this chapter from Oliver Wendell Homes implies that the idea of critical thinking will expand our mind, and that seems desirable. But, he also cautions us with this thought:

Insanity is often the logic of an ***accurate*** *mind* ***overtasked.***[5]

This points out an apparent paradox. Becoming hypersensitive to seeking evidence that is of sterling quality and using reasoning that is beyond reproach can leave us mentally exhausted. Although critical thinking can help ensure that we are more accurate in our conclusions, having an obsession with critical thinking can tax our sanity to the limit.

We seek accuracy in our thoughts and nutrition in our food. But, can we become so fixated on eating fruits and vegetables that we never indulge in a sweet and fattening dessert? The dessert provides pleasure, even though we know it is generally bad for the body. Similarly, should we force our brains to consume only the peas and carrots of critical thinking? Will such a logical diet make us emotionally depressed?

Interestingly, this is actually a notion that has gained a fair amount of attention in social psychology for decades. Some evidence indicates that those who are a bit more pessimistic than average actually have a better grip on reality than their more optimistic cousins.[6] They know exactly how frequently most negative life events occur—illness, pain, accidents, job loss, divorce, and death. They can see what some optimists refuse to see—the corruption, greed, poverty, starvation, violence, murder, repression, genocide, terrorism, and suffering. In fact, this phenomenon is known as *depressive realism (also called the sadder but wiser effect),* and it suggests that those people who are a bit gloomier than average often have a more accurate view of reality. It is not that they obsessively focus on the negative. In fact, unlike many individuals, they are willing to look at ***both*** the positive ***and*** the negative.

This *sadder but wiser effect* does not directly contradict the dominant psychological theory that many people who are extremely and clinically depressed suffer from excessive negative thinking and fatalistic distortions of reality. Many clinically depressed people, for example, get into a cycle of thinking in which they can conceive of only the worst possible consequences. Counseling and drug therapies designed to change this cycle of negative thinking have helped untold numbers of depressed individuals manage their conditions.

As used here, *depressive realism* does ***not*** refer being clinically depressed. It refers to people who are a bit more solemn and somber than the average or normal person. As mentioned before, these somber individuals often see the world much more accurately than do the masses of people. Is it their accuracy—and willingness to look at both the positive and the negative—that is responsible for inducing a less than optimistic attitude? Or, could it be that an initial pessimistic individual is more likely to seek the whole truth—the positive and the negative? Or are there other notions that help explain why research finds this *sadder but wiser* phenomenon?

Susan C. Vaughan, M.D., author of *Half Empty Half Full: Understanding the Psychological Roots of Optimism,* states: *The sadder but wiser* individuals tend to have more realistic perceptions of how much control they have over certain events.

> *Their perceptions of themselves are closer to reality than [those of] their more positive-thinking counterparts. They have a better understanding of how they compare with other people. They are also more accurate at estimating the risks of all sorts of disasters, . . . from plane crashes to their chances of being hit by a bus when crossing the street on any given day. They saw the dangers of life head-on and estimated them accurately.*[7]

In other words, unlike many people, they were more accurate on subjective measures and objective measures. They had a better understanding, for example, of how they compared athletically with others. But they also had a better understanding of objective reality. For example, as mentioned above, they could estimate all types of risks with far more accuracy than normal people could. They were less likely to commit many of the thinking and perceptual errors listed in this chapter and throughout the book.

To be more specific, millions of college students taking social psychology courses are familiar with the standard summary of *depressive realism* as provided by Shelly Taylor, Ph.D., a prominent researcher and theorist. One should note that Taylor is ***not*** using the word *depression* to signify *clinical depression,* but rather as meaning being in a mental state that is a bit more reflective, serious, and somber than most. She also is not necessarily suggesting that *normal* refers to *ideal.* Being average, or at the norm (normal), is hardly a cause for celebration for most people. One norm is that most people do not exercise enough; this *normal* behavior is *not ideal* behavior. Given these definitions, consider her summary.

> *Normal people exaggerate how competent and well liked they are. Depressed people do not. Normal people remember their past behavior with a rosy glow. Depressed people (unless severely depressed) are more evenhanded in recalling their success and failures. Normal people describe themselves primarily positively. Depressed people describe both their positive and negative qualities. Normal people take credit for successful outcomes and tend to deny responsibility for failure. Depressed people accept responsibility for both success and failure. Normal people exaggerate the control they have over what goes on around them. Depressed people are less vulnerable to the illusion of control. Normal people believe to an unrealistic degree that the future holds a bounty of good things and few bad things. Depressed people are more realistic in their perceptions of the future. In fact, on virtually every point on which normal people show enhanced self-regard, illusions of control, and unrealistic visions of the future, depressed people fail to show the same biases.*[8]

In addition, most normal people believe that others pay more attention to them than is actually the case. More somber people correctly realize that most people are so busy with their own problems and lives that they rarely have any remaining time to worry very much about the lives of others.

In high school we worried that *everyone* would stare at our *enormous* new acne blemish. However, the reality is that few, if anyone noticed. It did ***not*** become the *talk of the town*. Any embarrassment was caused by our internal distorted thoughts, not by external reactions we evoked from others.

This delusion of self-importance from high school often stays with people their entire lives. More somber people, for example, realize that they could dye their hair and most people would not notice. And those who did notice would pay attention for a few minutes and then go back to their own lives. Somber people realize that, regardless of their problems and tragedies, the world keeps turning as if nothing had happened. In other words, they do not magnify their own self-importance.

You think too much is a criticism often directed at these slightly more somber, introspective people. And frequently the comments come from those that are suspiciously too happy—or don't think quite enough, perhaps free spirits with delusional thoughts. Such people have made friends with a number of comforting notions:

Whatever will be, will be. . . .

It was never meant to be. . . .

It's fate. . . .

It's destiny. . . .

It is not our job to figure out why. . . .

Don't try to understand, just accept it. . . .

Are these comforting thoughts grounded in reality? Suppose both a serious-minded person and a happy-go-lucky, free spirit apply for a job, and neither is hired. Upon being rejected for the job, the serious one is apt to be a bit depressed—he takes at least some responsibility for the outcome. But, the free spirit can avoid this notion of personal responsibility by simply calling up that comforting thought: *It just wasn't meant to be.* Furthermore, he is much more likely to rationalize the outcome with another friendly notion: *I didn't want that stupid job anyway.*

Naive people are often very loveable and free spirits are frequently envied. But, could it be that these individuals are actually the *smart ones?* After all, if the goal of life is happiness, it seems as if they had achieved their goal. Should we allow these individuals to live in the playgrounds of their slightly delusional minds?

Positive Illusions

Is it our responsibility to encourage people to become more realistic, at the risk of denting their optimism? Or, is it best to let people pursue comfort in their lives regardless of their methods. Should we make people see reality at the risk of making them less happy? Or, instead, should the realists just drink a bit more from the bottle of delusion, enjoy life, and sing: *Don't Worry—Be Happy?*

On a personal note about enjoying life, once a month my wife and I go bowling with friends, and occasionally I forget to bring my glasses. When I have my glasses, I can look all the way down the long bowling alley and clearly see all ten individual pins. Almost subconsciously, I begin to think of how difficult it is going to be to knock down all ten pins. Doubt has seeped into my mind. However, when I forget my glasses, my vision is so poor that I do not actually see all ten individual pins. Instead, I see one white solid object, and suddenly I feel my sense of confidence rising—all I have to do is hit that large object. It is this illusion that makes me temporarily more optimistic. Should we consciously create these mirages as a way of making our lives more enjoyable?

One needs only to visit a bar to see that some individuals consciously seek to create delusions on a regular basis. Some will acknowledge that the goal of *drinking* is to get *drunk.* This intention is stated prior to the first drink. In other words, rational people are consciously electing to put themselves into a state where they know they will become somewhat delusional and irrational. The intent is to see other people and the world in a way that is more comforting and enjoyable. Fortunately, most people realize that the temporary state of extreme happiness or euphoria is usually paid for by a future period of pain—the next morning's hangover.

Obviously, alcohol and drugs are not tools to build long-term happiness or pleasure. However, is there a less toxic way of inducing comforting illusions?

In discussing *depressive realism,* both Vaughan and Taylor identified the many ways that *normal* people are often out of touch with reality. Despite the seemingly negative picture they paint of most *normal* people, they do ***not*** suggest that normal people need to become more serious, somber, and realistic. In fact, and surprisingly, both argue just the opposite.

In Taylor's book, *Positive Illusions: Creative Self-Deception and the Healthy Mind (1991),*[9] she explains how and why those who are a bit more somber must learn to create illusions and delusions to increase happiness. She doesn't suggest that they should ignore their accurate perceptions of the world. Instead, she encourages such people to consciously limit their perceptions of negative events.

Leading psychiatrist, psychoanalyst, and researcher, Dr. Susan C. Vaughan, agrees. She says that developing optimism depends upon using methods to fool the brain's circuitry into looking on the bright side of life. In her book, *Half Empty, Half Full,* she explains how one creates delusions in life to become happier.[10]

Likewise, Daniel Goleman, Ph.D., internationally known author of *Emotional Intelligence* (1995), psychologist, and brain functioning expert, provides extensive evidence showing that people often dim their own awareness as a way of preventing uncomfortable thoughts and feelings.[11] That is, whether consciously or not, many people make delusions and illusions their daily companions.

In the popular movie, *A Few Good Men,* actor Jack Nicholson plays a Navy general who is being grilled in a courtroom by an attorney played by Tom

Cruise. The goal of the cross-examination is to anger the general to the point that he loses control—and admits that illegal and torturous tactics are used to train recruits to become loyal soldiers. In the heated exchange, the attorney demands, ***I want the truth!*** Of course, it is at this point that the agitated general declares, ***You can't handle the truth!*** Are there times when we can't handle the truth—or prefer to distort it—or prefer to ignore it—as a way of creating an artificial sense of optimism?

Clearly, it is easy to understand how one might feel rather somber and trivial being just one person in a world with billions of people. That world, of course, is one of millions of such planets in a massive universe that is billions of years old. Our lives, nothing more than drops in an ocean of vast time and space, can make us seem rather irrelevant and insignificant. However, perhaps a healthy delusion is to recall that popular quote: *You might be just one person in this world, but perhaps to one person you just might be the world.*[12]

Similarly, ignoring the piles of pessimistic evidence by focusing on the token signs of optimism is sometimes required to generate happiness. Most are familiar with the following story.

> *A young girl was walking on the beach early one morning and noticed that the receding tide had left thousands of starfish stranded on the sand. Rather than let them die, the girl started tossing them back into the sea. A nearby fisherman shook his head in dismay and shouted to her that it wouldn't matter because there were thousands of starfish. With that, she picked up a starfish, threw it as far into the sea as she could, and exclaimed, it matters to that one!* [13]

Final Thought

It is certainty true that critical thinking asks us to question much of what we think and believe. Sometimes our questioning leads us to answers that are troublesome and uncomfortable. As mentioned in a previous chapter, *one goal of education is to provide answers, but another more important goal is to question the answers.*

Madsen Pirie writes:

> *Intelligent people are often afraid of being thought rather cold because they use reason so much. They do not want to appear to be emotionally deficient; . . . they happily abandon reason as the price of their admission ticket into the human race.*[14]

Pirie's comments are worth noting, but those who are worried that intellect will suffocate emotion and optimism should consider that using knowledge and reasoning is a form of caring. It has been said, *people don't really care* ***how much you know . . . they just want to know how much you care.*** I would suggest that most people, who truly care, also take the time and energy to ensure that they really know—that their knowledge is based on evidence and reasoning, not on delusion or illusion. A caring doctor, for example, is not simply one who shows compassion and seems to have a pleasant bedside manner. More importantly, a doctor who genuinely cares ensures that his or her information is accurate and reasoning valid. Diagnosing the symptoms correctly and prescribing the most effective treatment known demonstrates the highest possible level of caring and compassion.

In closing, we need to acknowledge that intellect is not divorced from emotion and optimism. In fact, it is the passion of the heart that fuels the engine of the mind.

Notes

Chapter One

[1] Quotations for Creative Thinking, "Creative Quotations from Burton Hillis," http://creativequotations.com/one/2313.htm (accessed June 2, 2009).

[2] The Quotations Page, "Adolf Hitler: What luck for rulers..." http://www.quotationspage.com/quote/971.html (accessed June 2, 2009).

[3] ThinkExist.com Quotations, "Ayn Rand quotes," http://thinkexist.com/quotation/wealth_is_the_product_of_mans_capacity_to_think/220796.html (accessed June 2, 2009).

Chapter Two

[1] MSN Encarta : Online Dictionary, s.v. "logic," http://encarta.msn.com/dictionary_1861699377/logic.html (accessed June 2, 2009).

[2] W. Ward Holther and W. B. Fearnside, *Fallacy: The Counterfeit of Argument* (New York: Spectrum, 1959).

[3] Douglas N. Walton, *Plausible Argument in Everyday Conversation* (S U N Y Series in Speech Communication) (Albany, NY: State University of New York Press, 1992), 268.

Chapter Three

[1] ThinkExist.com Quotations, "Mark Twain Quotes," http://thinkexist.com/quotation/all_generalizations_are_false-including_this_one/216596.html (accessed June 2, 2009).

[2] Malcolm W. Browne, "Physicists Debunk Claim of a New Kind of Fusion," The New York Times, http://partners.nytimes.com/library/national/science/050399sci-cold-fusion.html (accessed June 2, 2009).

[3] "The Accuracy of National Pre-Election Polls," Oxford Journals (Social Sciences) Public Opinion Quarterly, http://poq.oxfordjournals.org/cgi/content/full/69/5/642 (accessed June 2, 2009).

[4] "10th Anniversary of Columbine Massacre," Naples Daily News, http://www.naplesnews.com/photos/galleries/2009/apr/19/ten-years-later-columbines-hold-remains-strong/ (accessed June 2, 2009).

[5] "Texas Bills Would Allow Guns at College," msnbc.com (Education). http://www.msnbc.msn.com/id/29944438/ (accessed June 2, 2009).

[6] US Census Bureau, "US Census Press Releases," Census Bureau Home Page, http://www.census.gov/Press-Release/www/releases/archives/facts_for_features_special_editions/007108.html (accessed June 2, 2009).

[7] Matthew Harwood, "Murders at Schools Drop," Security Management Homepage, http://www.securitymanagement.com/news/murders-schools-drop (accessed June 2, 2009).

[8] "Heart Disease Statistics." MamasHealth.com, http://www.mamashealth.com/HEART_stat.asp (accessed June 2, 2009).

[9] Mary Carter, "Heart disease still the most likely reason you'll die," CNN.com, http://www.cnn.com/2006/HEALTH/10/30/heart.overview/index.html (accessed June 2, 2009).

[10] American Cancer Society, "Cigarette Smoking," American Cancer Society Homepage, http://www.cancer.org/docroot/PED/content/PED_10_2X_Cigarette_Smoking.asp (accessed June 12, 2009).

[11] "California ban on same-sex marriage struck down," CNN.com, http://www.cnn.com/2008/US/05/15/same.sex.marriage/ (accessed June 2, 2009).

[12] "The Family: Interracial Marriages Are Becoming More Common," Social Issues Reference, http://social.jrank.org/pages/889/Family-Interracial-Marriages-Are-Becoming-More-Common.html (accessed June 2, 2009).

[13] Amy Teeple, "BMI Calculator , Body Mass Index Chart , Ideal Body Weight Chart Review," DocShop.com, http://www.docshop.com/2008/04/08/arnold-schwarzenegger-is-obese-problems-with-body-mass-index-bmi-calculations/ (accessed June 2, 2009).

[14] Robert J. Gula, *Nonsense: A Handbook of Logical Fallacies*. (New York: Axios Press, 2002), 98.

Chapter Four

[1] Christianpost.com, "Georgia Gets Rain Day After Gov. Leads Prayers to God," http://www.christianpost.com/article/20071115/georgia-gets-rain-day-after-gov-leads-prayers-to-god/index.html (accessed June 2, 2009).

[2] Sympatico, "Paradigms," http://www3.sympatico.ca/saburns/pg0203.htm (accessed June 2, 2009).

[3] Francis Collins, "A New Creationist Take on Darwinism," Arizona Republic Newspaper, May 12, 2009, p. B11.

[4] The Fallacy Files, "Logical Fallacy: Cum Hoc Ergo Propter Hoc," http://www.fallacyfiles.org/cumhocfa.html (accessed June 2, 2009).

[5] Claudio Sanchez, "SAT Math Scores Peak; Verbal Remains Flat," National Public Radio (NPR), http://www.npr.org/templates/story/story.php?storyId=4824559 (accessed June 2, 2009).

[6] Inside Higher Ed, "SAT Math Scores Are Up," http://www.insidehighered.com/news/2005/08/31/sat (accessed June 2, 2009).

[7] Logical Fallacies, "A Post Hoc Fallacy," http://www.logicalfallacies.info/presumption/post-hoc/ (accessed June 2, 2009).

[8] C. K. Roberts, "Weight Loss with a Low-Carbohydrate, Mediterranean, or Low-Fat Diet," The New England Journal of Medicine, http://content.nejm.org/cgi/content/full/359/3/229 (accessed June 2, 2009).

[9] FOXNews.com , "Study: Low-Carb Diet Best for Weight, Cholesterol," http://www.foxnews.com/story/0,2933,384362,00.html (accessed June 2, 2009).

[10] "Depression and Alcohol," The Alcoholism Home Page, http://alcoholism.about.com/od/depress/Depression_and_Alcohol.htm (accessed June 2, 2009).

[11] WebMD, "Drink More Diet Soda, Gain More Weight?" http://www.webmd.com/diet/news/20050613/drink-more-diet-soda-gain-more-weight (accessed June 2, 2009).

[12] Surgery.org, "Facelift: Does Looking Younger Help You Live Longer?" http://www.surgery.org/press/news-release.php?iid=168 (accessed June 2, 2009).

[13] Internet Encyclopedia of Philosophy (IEP), s.v., "Fallacies," http://www.iep.utm.edu/f/fallacy.htm#Common%20Cause (accessed June 13, 2009).

[14] US News & World Report, "How Much is That College Degree Really Worth?" 30 October 2008, http://www.usnews.com/articles/education/2008/10/30/how-much-is-that-college-degree-really-worth.html (accessed June 3, 2009).

[15] Penelope Wang, "The real return on your college investment," CNNMoney.com, http://money.cnn.com/2008/08/20/pf/college/college_price.moneymag/index3.htm (accessed June 3, 2009).

[16] Peter Gray, "Reasons to Consider a Less Selective, Less Expensive College: Saving Money is Just One of Them," Psychology Today, 28 June 2009, http://www.psychologytoday.com/blog/freedom-learn/200810/reasons-consider-less-selective-less-expensive-college-saving-money-is-jus (accessed July 1, 2009).

[17] Patrick J. Hurley, *A Concise Introduction to Logic* (Belmont, CA: Thomson Wadsworth, 2007), 154.

[18] The Skeptic's Dictionary, "Texas sharpshooter fallacy," http://skepdic.com/texas.html (accessed June 3, 2009).

[19] Logical Fallacies, "The Texas Sharpshooter Fallacy," The Fallacy Files, http://www.fallacyfiles.org/texsharp.html (accessed June 3, 2009).

[20] Thomas Gilovich, *How We Know What Isn't So: The Fallibility of Human Reason in Everyday Life* (New York: Free Press, 1993), 11–17.

[21] Sports Illustrated, "Alex Wolff: Unraveling the SI cover jinx," 29 August 2003, http://sportsillustrated.cnn.com/2003/writers/alexander_wolff/08/27/viewpoint_flashback/index.html (accessed June 3, 2009).

[22] Tim Smith, –"Millions of superstitious readers-- and many athletes-- believe that an appearance on Sports Illustrated's cover is the kiss of death. But is there really such a thing as the SI Jinx? 27 August 2003, http://sportsillustrated.cnn.com/2003/magazine/08/27/jinx/ (accessed June 3, 2009).

[23] David G. Myers, "*Intuition: Its Powers and Perils (Yale Nota Bene)* (New Haven: Yale University Press, 2004), 146–148.

[24] GeoCities, "Irrational Exuberance Speech: Fed Chairman Alan Greenspan," http://www.geocities.com/ecocorner/intelarea/ag1.html (accessed June 3, 2009).

[25] Fidelity Monitor, "Following Our Models," http://www.fidelitymonitor.com/followingmodels.php (accessed June 3, 2009).

[26] James K. Glassman, "Selling Short on NRO Financial," National Review Online, http://www.nationalreview.com/nrof_glassman/glassman032602.asp (accessed June 3, 2009).

[27] Madsen Pirie, *How to Win Every Argument: The Use and Abuse of Logic* (New York: Continuum International Publishing Group, 2006), 69.

[28] Maureen Stoss,. Email Interview, 22 June 2009.

[29] "Bingo and Online Slot Machines," http://www.bingo.cx/slots.html (accessed June 3, 2009).

[30] BrainyQuote, "Arthur C. Clarke Quotes," http://www.brainyquote.com/quotes/quotes/a/arthurccl166777.html (accessed June 3, 2009).

[31] Paranormality.com, "Barnum Effect," http://www.paranormality.com/barnum_effect.shtml (accessed June 3, 2009).

[32] Thomas Gilovich, *How We Know What Isn't So: The Fallibility of Human Reason in Everyday Life* (New York : Free Press, 1993), 62.

[33] Howard Kahane, *Logic and Contemporary Rhetoric: The Use of Reason in Everyday Life* (Belmont, CA: Wadsworth, 1976), 137.

[34] Howard Kahane, *Logic and Contemporary Rhetoric: The Use of Reason in Everyday Life* (Belmont, CA: Wadsworth, 1976), 136.

[35] James Randi Educational Foundation, "Million Dollar Challenge FAQ," http://www.randi.org/site/index.php/1m-challenge/challenge-faq.html (accessed June 3, 2009).

[36] James Randi Educational Foundation, "Challenge Info," http://www.randi.org/site/index.php/1m-challenge.html (accessed June 3, 2009).

[37] James Randi Educational Foundation, "Challenge Applications," JREF Forum, http://forums.randi.org/forumdisplay.php?f=43&sort=lastpost&order=desc&page=6 (accessed June 13, 2009).

[38] Christian Apologetics & Research Ministry, "Extraordinary Claims Require Extraordinary Evidence, http://www.carm.org/evidence/extraordinary.htm (accessed June 3, 2009).

Chapter Five

[1] Montecito Journal, "Ward Connerly," http://www.montecitojournal.net/archive/13/7/758/ (accessed June 3, 2009).

[2] Holocaust Educational Resource, "Fallacy: Ad Hominem," http://www.nizkor.org/features/fallacies/ad-hominem.html (accessed June 3, 2009).

[3] David M. Herszenhorn, "Estimates of Iraq War Cost Were Not Close to Ballpark" New York Times Advertisement, 19 March 2008, http://www.nytimes.com/2008/03/19/washington/19cost.html (accessed June 3, 2009).

[4] Ken Dilanian, "War Costs May Total $2.4 Trillion," USATODAY.com, 28 June 2009, http://www.usatoday.com/news/military/2007-10-23-wacosts_N.htm (accessed July 1, 2009).

[5] Three Trillion Dollar War: The True Cost of the Iraq Conflict, http://threetrilliondollarwar.org/ (accessed June 3, 2009).

[6] Ken Dilanian, "War Costs May Total $2.4 Trillion," USATODAY.com, 28 June 2009, http://www.usatoday.com/news/military/2007-10-23-wacosts_N.htm (accessed July 1, 2009).

[7] MSN.com, "Clinton offers universal health care plan," http://www.msnbc.msn.com/id/20819827/ (accessed June 3, 2009).

[8] David Welna, "McCain and Bush: Common Ground on War, Torture,"National Public Radio (NPR), http://www.npr.org/templates/story/story.php?storyId=91560330 (accessed June 3, 2009).

[9] Holocaust Educational Resource, "Fallacy: Ad Hominem Tu Quoque," http://www.nizkor.org/features/fallacies/ad-hominem-tu-quoque.html (accessed June 3, 2009).

[10] USATODAY.com, "Statements on Atkins' death," http://www.usatoday.com/news/health/2004-02-10-atkins-statements_x.htm (accessed June 3, 2009).

[11] Answers.com—Online Dictionary, s.v. "big lie," http://www.answers.com/topic/big-lie (accessed June 4, 2009).

[12] Richard Leiby, "A Couch Tom Cruise Won't Jump On," Washingtonpost.com, http://www.washingtonpost.com/wp-dyn/content/article/2005/06/24/AR2005062401866.html (accessed June 3, 2009).

[13] Jeannette Walls Pearson, "Psychiatrists Think Cruise Should Have Head Examined," The Ross Institute Internet Archives for the Study of Destructive Cults, http://www.rickross.com/reference/scientology/celebrities/celebrities39.html (accessed June 4, 2009).

[14] Austin Cline, "Tom Cruise vs. Mental Health Experts," http://atheism.about.com/b/2005/07/07/tom-cruise-vs-mental-health-experts.htm (accessed June 4, 2009).

Chapter Six

[1] ThinkExist.com, "Mark Twain quotes," http://thinkexist.com/quotation/the_difference_between_the_right_word_and_the/11504.html (accessed June 4, 2009).

[2] Brian Dakss, "Pregnant Woman Loses HOV Lane Case," CBS News, http://www.cbsnews.com/stories/2006/01/12/earlyshow/main1203514.shtml (accessed June 4, 2009).

[3] Dennis Overbye, "Philosophers Notwithstanding, Kansas School Board Redefines Science," The New York Times, 15 November 2005, http://www.nytimes.com/2005/11/15/science/sciencespecial2/15evol.html?pagewanted=print (accessed June 4, 2009).

[4] LegalDaily.com, "U.S. Legal Memorandum on Torture of Terror Suspects and Detainees," http://www.legaldaily.com/ (accessed June 13, 2009).

[5] United States Department of Justice, "Legal Standards Applicable Under 18 U.S.C. §§ 2340-2340A,". http://www.usdoj.gov/olc/18us (accessed June 28, 2009).

[6] MSN.com, "Bush: U.S.A. Does Not Torture People," http://www.msnbc.msn.com/id/21148801/ (accessed June 4, 2009).

[7] Los Angeles Times, "Waterboarding is Legal, White House Says," http://www.latimes.com/news/nationworld/nation/la-na-torture7feb07,1,3156438.story (accessed June 4, 2009).

[8] Steven D. Schwinn, Ruthann Robson, and Nareissa L. Smith, "White House to Release Details of Enhanced Interrogation Techniques," USLaw.com, http://www.uslaw.com/library/Constitutional_Law/White_House_Release_Details_Enhanced_Interrogation_Techniques.php?item=422110 (accessed June 4, 2009).

[9] Alfred W. McCoy, "A Question of Torture: CIA," Google Book Search, http://books.google.com/books?id=-sa-nUI3HGMC&pg=PA113&lpg=PA113&dq=Bush:+%22illegal+enemy+combatants%22&source=bl&ots (accessed June 4, 2009).

[10] Ablogistan.com, "Hillary Clinton wasn't misled about the Iraq war vote," http://www.ablogistan.com/archives/2008/01/people_died_cli.html (accessed June 4, 2009).

[11] William J. Bernstein, "Stock Picking is a Losing Proposition, CNN.com, http://money.cnn.com/2009/05/09/magazines/moneymag/stock-strategies.moneymag/ (accessed June 28, 2009).

[12] The Huffington Post, "McCain On Obama: Clear Who Hamas Wants to be the Next President," http://www.huffingtonpost.com/2008/04/25/mccain-on-obama-clear-who_n_98614.html (accessed June 24, 2009).

[13] Charles Dharapak, "Obama: McCain Lost his Bearings with Hamas Comment," USATODAY.com, http://www.usatoday.com/news/politics/election2008/2008-05-08-obama-mccain_N.htm (accessed June 24, 2009).

[14] McCain: Not Offended By Alleged Age-Baiting: Political Radar," ABC News, http://blogs.abcnews.com/politicalradar/2008/05/mccain-not-offe.html (accessed June 24, 2009).

[15] Truman Library, "Clark M. Clifford Oral History Interview, February 14, 1973," http://www.trumanlibrary.org/oralhist/cliford7.htm (accessed June 24, 2009).

[16] The Quotations Page, "Moses Hadas Quotes," http://www.quotationspage.com/quotes/Moses_Hadas/ (accessed June 24, 2009).

[17] Office-humour.co.uk, "English professor writes..." http://office-humour.co.uk/item/6312/ (accessed June 24, 2009).

[18] RinkWorks, "Things People Said: Car Accident Reports," http://www.rinkworks.com/said/insurance.shtml (accessed June 24, 2009).

[19] "Snopes.com, "Welfare Application Humor," http://www.snopes.com/humor/lists/welfare.asp (accessed June 24, 2009).

Chapter Seven

[1] Austin Cline, "Gullibility 101: What is Gullibility? What Sorts of Beliefs are Produced by Gullibility and by Being Gullible?" http://atheism.about.com/od/aboutskepticism/p/Gullibility.htm (accessed June 4, 2009).

[2] BrainyQuote, "William Cowper Quotes," http://www.brainyquote.com/quotes/authors/w/william_cowper.html (accessed June 4, 2009).

[3] Media Partners, "Groupthink video," http://www.media-partners.com/communication_skills/groupthink.htm (accessed June 4, 2009).

[4] Media Partners, "Groupthink video," http://www.media-partners.com/communication_skills/groupthink.htm (accessed June 4, 2009).

[5] Lauren Johnston, "CIA Faulted For Iraq Flaws," CBS News, http://www.cbsnews.com/stories/2004/07/09/iraq/main628550.shtml (accessed June 4, 2009).

[6] Steve Denning, "Why giving people reasons does not work," http://www.stevedenning.com/Documents/WhyGivingReasonsDoesntWork.html (accessed June 4, 2009).

[7] Steve Denning, "Why giving people reasons does not work," http://www.stevedenning.com/Documents/WhyGivingReasonsDoesntWork.html (accessed June 4, 2009).

[8] Time magazine, "Can't Make Up Your Mind?" 20 October 2008.

[9] Time magazine, "What Kind of Voter Are You?" 20 October 2008.

[10] Thomas Gilovich, *How We Know What Isn't So: The Fallibility of Human Reason in Everyday Life* (New York: Free Press, 1993).

[11] Mary Shaw, "Bush Admits He Would Have Gone to War Knowing What We Know Now," Associated Content, http://www.associatedcontent.com/article/17140/bush_admits_he_would_have_gone_to_war.html (accessed June 24, 2009).

[12] Michael C. Dorf, "Kerry Stands by His Iraq War Vote," http://writ.news.findlaw.com/dorf/20040818.html (accessed June 24, 2009).

[13] David Gollust, "US Envoy Will Join in Iran Nuclear Meeting," Voice of America, http://www.voanews.com/english/archive/2008-07/2008-07-16-voa6.cfm?CFID=230719385&CFTOKEN=47842336&jsessionid=00308408fe5bc3a84ef87d7ea2c46e10604e (accessed June 24, 2009).

[14] Ed Hornick and Alexander Marquardt, "Obama says offshore drilling stance nothing new," CNN.com, http://politicalticker.blogs.cnn.com/2008/08/02/obama-defends-offshore-oil-drilling-policy-change/ (accessed June 24, 2009).

[15] ThinkExist.com, "Groucho Marx quotes," http://thinkexist.com/quotation/i_refuse_to_join_any_club_that_would_have_me_as_a/207365.html (accessed June 24, 2009).

[16] Mary Bustamante, and Renee Schafer Horton, "Bill Would Ban Anti-Democratic Teaching," Tucson Citizen, http://www.tucsoncitizen.com/daily/local/82833.php (accessed June 24, 2009).

[17] Austin Hill, "Prop. 105: Good Intent, Bad Results," EastValleyTribune.com, http://www.eastvalleytribune.com/story/127667 (accessed June 4, 2009).

[18] Robert Robb, "Vote NO on Prop 105," http://thevotersofaz.com/?cat=4 (accessed June 24, 2009).

[19] ThinkQuest Library, "Tiger Woods," http://library.thinkquest.org/5672/Tiger_Woods.html (accessed June 4, 2009).

[20] Consumer Reports, "New car reviews and buying advice from Consumer Reports," http://www.consumerreports.org/cro/cars/new-cars/ (accessed June 4, 2009).

[21] CNN.com, "Bush Insists Iraq, al Qaeda had 'relationship,'" 17 June 2004, http://www.cnn.com/2004/ALLPOLITICS/06/17/Bush.alqaeda/ (accessed June 24, 2009).

Chapter Eight

[1] Jokes Duniya, "Lipstick prints on mirror," http://www.jokesduniya.com/3916/lipstick-prints-on-mirror/ (accessed June 4, 2009).

[2] Goodreads.com, "Quotes by Desmond Tutu," http://www.goodreads.com/author/quotes/5943.Desmond_Tutu (accessed June 24, 2009).

[3] Stephanie M. Skier, "Polaroid Corporation Files for Bankruptcy," The Harvard Crimson, http://www.thecrimson.com/article.aspx?ref=121699 (accessed June 24, 2009).

[4] Claudia H. Deutsch, "Kodak to Lay Off 10,000 Employees in a 10% Cutback," The New York Times, 12 November 1997, http://www.nytimes.com/1997/11/12/business/kodak-to-lay-off-10000-employees-in-a-10-cutback.html?pagewanted=all (accessed June 24, 2009).

[5] William Branigin and Dana Priest, " Senate Report Blasts Intelligence Agencies' Flaws," Washingtonpost.com, http://www.washingtonpost.com/ac2/wp-dyn/A38459-2004Jul9?language=printer (accessed June 4, 2009).

[6] Charles Brown, "When Can I Appeal A Court's Decision?" EzineArticles, http://ezinearticles.com/?When-Can-I-Appeal-A-Courts-Decision?&id=444462 (accessed June 30, 2009).

[7] U.S. Census Bureau, "State Ranking--Statistical Abstract of the United States," www.census.gov/statab/ranks/rank01.html (accessed June 24, 2009).

[8] Gary Nine, "Respect for Regular Folks: Arizona K–12 Arizona Education Points," http://www.respectforregularfolks.com/index.php/home/34-content/115-arizona-k-12-arizona-education-points (accessed June 25, 2009).

[9]John Faherty, "Arizona ranked dumbest in U.S.," Arizona Republic, http://www.azcentral.com/arizonarepublic/news/articles/1018dumb1018.html (accessed June 25, 2009).

[10]The Disaster Center, "United States Crime Rates 1960–2007," http://www.disastercenter.com/crime/uscrime.htm (accessed June 25, 2009).

[11]Michigan.gov, "Pop Density-Land Area 24 Cities," http://74.125.95.132/search?q=cache:TUuwlc0Xb3YJ:www.michigan.gov/documents/hal_lm_census_landarea (accessed June 25, 2009).

[12]CelebrityGenius.com "Ted Turner Gossip, Pictures, Videos, Biography and More," http://www.celebritygenius.com/Ted-Turner (accessed June 25, 2009).

[13]Goodreads, "Quotes by Ronald Reagan," http://www.goodreads.com/author/quotes/3543.Ronald_Reagan (accessed June 25, 2009).

Chapter Nine

[1]Quotatio, "Oliver Wendell Holmes Quotations: Index 4," http://www.quotatio.com/h/holmes-oliver-wendell-quotes.html (accessed June 4, 2009).

[2]Wikipedia, s.v. "list of cognitive biases," http://en.wikipedia.org/wiki/List_of_cognitive_biases (accessed June 4, 2009).

[3]Thomas R. Hoerr, *Becoming a Multiple Intelligences School* (Alexandria, VA: Association For Supervision & Curriculum Development, 2000).

[4]Quotations Page, " Quote Details: Father Larry Lorenzoni: The Average Person Thinks," http://www.quotationspage.com/quote/993.html (accessed June 4, 2009).

[5]BrainyQuote, "Oliver Wendell Holmes Quotes," http://www.brainyquote.com/quotes/quotes/o/oliverwend152694.html (accessed June 4, 2009).

[6]Susan C. Vaughan, M.D., *Half Empty Half Full: Understanding the Psychological Roots of Optimism* (New York: Harcourt, 2000), 7.

[7]Susan C. Vaughan, M.D., *Half Empty Half Full: Understanding the Psychological Roots of Optimism* (New York: Harcourt, 2000), 6–7.

[8]David Myers, *Social Psychology*, 8th ed. (New York: McGraw-Hill, 2005), 577.

[9]Shelly E. Taylor, "Positive Illusions: Creative Self-Deception and the Healthy Mind," http://www.amazon.com/Positive-Illusions-Creative-Self-Deception-Healthy/dp/0465060536 (accessed June 25, 2009).

[10]Susan C. Vaughan, M.D., *Half Empty Half Full: Understanding the Psychological Roots of Optimism* (New York: Harcourt, 2000), xiv.

[11] Daniel Goleman,. *Emotional Intelligence: Why It Can Matter More Than IQ* (New York: Bantam, 1995).

[12] ThinkExist.com, "Brandi Snyder Quotes," http://thinkexist.com/quotation/to_the_world_you_may_be_just_one_person-but_to/189999.html (accessed June 25, 2009).

[13] AccessMyLibrary, "The Starfish Story: Forget Trying to Fix All of School's Problems, Concentrate on One Student at a Time, http://www.accessmylibrary.com/coms2/summary_0286-3389794_ITM (accessed June 25, 2009).

[14] Madsen Pirie, *How to Win Every Argument: The Use and Abuse of Logic* (New York: Continuum International Publishing Group, 2007), 57.

Bibliography

"10th Anniversary of Columbine Massacre." Naples Daily News. http://www.naplesnews.com/photos/galleries/2009/apr/19/ten-years-later-columbines-hold-remains-strong/ (accessed June 2, 2009).

Ablogistan.com. "Hillary Clinton wasn't misled about the Iraq war vote." http://www.ablogistan.com/archives/2008/01/people_died_cli.html (accessed June 4, 2009).

AccessMyLibrary, "The Starfish Story: Forget Trying to Fix All of School's Problems, Concentrate on One Student at a Time." http://www.accessmylibrary.com/coms2/summary_0286-3389794_ITM (accessed June 25, 2009).

"The Accuracy of National Pre-Election Polls." Oxford Journals (Social Sciences) Public Opinion Quarterly. http://poq.oxfordjournals.org/cgi/content/full/69/5/642 (accessed June 2, 2009).

American Cancer Society. "Cigarette Smoking." American Cancer Society Homepage. http://www.cancer.org/docroot/PED/content/PED_10_2X_Cigarette_Smoking.asp (accessed June 12, 2009).

Answers.com—Online Dictionary, s.v. "big lie." http://www.answers.com/topic/big-lie (accessed June 4, 2009).

Bernstein, William J. "Stock Picking is a Losing Proposition. CNN.com. http://money cnn.com/2009/05/09/magazines/moneymag/stock-strategies.moneymag/ (accessed June 28, 2009).

"Bingo and Online Slot Machines." http://www.bingo.cx/slots.html (accessed June 3, 2009).

BrainyQuote. "Arthur C. Clarke Quotes." http://www.brainyquote.com/quotes/quotes/a/arthurccl166777.html (accessed June 3, 2009).

BrainyQuote. "Oliver Wendell Holmes Quotes." http://www.brainyquote.com/quotes/quotes/o/oliverwend152694.html (accessed June 4, 2009).

BrainyQuote. "William Cowper Quotes." http://www.brainyquote.com/quotes/authors/w/william_cowper.html (accessed June 4, 2009).

Branigin, William, and Dana Priest. "Senate Report Blasts Intelligence Agencies' Flaws." Washingtonpost.com. http://www.washingtonpost.com/ac2/wp-dyn/A38459-2004Jul9?language=printer (accessed June 4, 2009).

Brown, Charles. "When Can I Appeal A Court's Decision?" EzineArticles. http://ezinearticles.com/?When-Can-I-Appeal-A-Courts-Decision?&id=444462 (accessed June 30, 2009).

Browne, Malcolm W. "Physicists Debunk Claim of a New Kind of Fusion." The New York Times. http://partners.nytimes.com/library/national/science/050399sci-cold-fusion.html (accessed June 2, 2009).

Bustamante, Mary, and Renee Schafer Horton. "Bill Would Ban Anti-Democratic Teaching." Tucson Citizen. http://www.tucsoncitizen.com/daily/local/82833.php (accessed June 24, 2009).

"California ban on same-sex marriage struck down." CNN.com. http://www.cnn.com/2008/US/05/15/same.sex.marriage/ (accessed June 2, 2009).

Carter, Mary. "Heart disease still the most likely reason you'll die." CNN.com. http://www.cnn.com/2006/HEALTH/10/30/heart.overview/index.html (accessed June 2, 2009).

CelebrityGenius.com. "Ted Turner Gossip, Pictures, Videos, Biography and More." http://www.celebritygenius.com/Ted-Turner (accessed June 25, 2009).

Christian Apologetics & Research Ministry. "Extraordinary Claims Require Extraordinary Evidence." http://www.carm.org/evidence/extraordinary.htm (accessed June 3, 2009).

Christianpost.com. "Georgia Gets Rain Day After Gov. Leads Prayers to God." http://www.christianpost.com/article/20071115/georgia-gets-rain-day-after-gov-leads-prayers-to-god/index.html (accessed June 2, 2009).

Cline, Austin. "Gullibility 101: What is Gullibility? What Sorts of Beliefs are Produced by Gullibility and by Being Gullible?" http://atheism.about.com/od/aboutskepticism/p/Gullibility.htm (accessed June 4, 2009).

Cline, Austin. "Tom Cruise vs. Mental Health Experts." http://atheism.about.com/b/2005/07/07/tom-cruise-vs-mental-health-experts.htm (accessed June 4, 2009).

CNN.com. "Bush Insists Iraq, al Qaeda had 'relationship,'" 17 June 2004. http://www.cnn.com/2004/ALLPOLITICS/06/17/Bush.alqaeda/ (accessed June 24, 2009).

Consumer Reports. "New Car Reviews and Buying Advice from Consumer Reports." http://www.consumerreports.org/cro/cars/new-cars/ (accessed June 4, 2009).

Dakss, Brian. "Pregnant Woman Loses HOV Lane Case." CBS News. http://www.cbsnews.com/stories/2006/01/12/earlyshow/main1203514.shtml (accessed June 4, 2009).

Denning, Steve. "Why Giving People Reasons Does Not Work." http://www.stevedenning.com/Documents/WhyGivingReasonsDoesntWork.html (accessed June 4, 2009).

Denning, Steve. "Why giving people reasons does not work." http://www.stevedenning.com/Documents/WhyGivingReasonsDoesntWork.html (accessed June 4, 2009).

"Depression and Alcohol." The Alcoholism Home Page. http://alcoholism.about.com/od/depress/Depression_and_Alcohol.htm (accessed June 2, 2009).

Deutsch, Claudia H. "Kodak to Lay Off 10,000 Employees in a 10% Cutback." The New York Times, 12 November 1997. http://www.nytimes.com/1997/11/12/business/kodak-to-lay-off-10000-employees-in-a-10-cutback.html?pagewanted=all (accessed June 24, 2009).

Dharapak, Charles. "Obama: McCain Lost His Bearings with Hamas Comment." USATODAY.com. http://www.usatoday.com/news/politics/election2008/2008-05-08-obama-mccain_N.htm (accessed June 24, 2009).

Dilanian, Ken. "War Costs May Total $2.4 Trillion." USATODAY.com, 28 June 2009. http://www.usatoday.com/news/military/2007-10-23-wacosts_N.htm (accessed July 1, 2009).

The Disaster Center. "United States Crime Rates 1960–2007." http://www.disastercenter.com/crime/uscrime.htm (accessed June 25, 2009).

Dorf, Michael C. "Kerry Stands by His Iraq War Vote." http://writ.news.findlaw.com/dorf/20040818.html (accessed June 24, 2009).

Faherty, John. "Arizona ranked dumbest in U.S." Arizona Republic. http://www.azcentral.com/arizonarepublic/news/articles/1018dumb1018.html (accessed June 25, 2009).

The Fallacy Files. "Logical Fallacy: Cum Hoc Ergo Propter Hoc." http://www.fallacyfiles.org/cumhocfa.html (accessed June 2, 2009).

"The Family: Interracial Marriages Are Becoming More Common." Social Issues Reference. http://social.jrank.org/pages/889/Family-Interracial-Marriages-Are-Becoming-More-Common.html (accessed June 2, 2009).

Fidelity Monitor. "Following Our Models." http://www.fidelitymonitor.com/followingmodels.php (accessed June 3, 2009).

FOXNews.com. "Study: Low-Carb Diet Best for Weight, Cholesterol." http://www.foxnews.com/story/0,2933,384362,00.html (accessed June 2, 2009).

Francis, Collins. "A New Creationist Take on Darwinism." Arizona Republic Newspaper, May 12, 2009, p. B11.

GeoCities. "Irrational Exuberance Speech: Fed Chairman Alan Greenspan." http://www.geocities.com/ecocorner/intelarea/ag1.html (accessed June 3, 2009).

Gilovich, Thomas. *How We Know What Isn't So: The Fallibility of Human Reason in Everyday Life.* New York : Free Press, 1993.

Glassman, James K. "Selling Short on NRO Financial." National Review Online. http://www.nationalreview.com/nrof_glassman/glassman032602.asp (accessed June 3, 2009).

Goleman, Daniel. *Emotional Intelligence: Why It Can Matter More Than IQ*. New York: Bantam, 1995.

Gollust, David. "US Envoy Will Join in Iran Nuclear Meeting." Voice of America. http://www.voanews.com/english/archive/2008-07/2008-07-16-voa6.cfm?CFID=230719385&CFTOKEN=47842336&jsessionid=00308408fe5bc3a84ef87d7ea2c46e10604e (accessed June 24, 2009).

Goodreads.com. "Quotes by Desmond Tutu." http://www.goodreads.com/author/quotes/5943.Desmond_Tutu (accessed June 24, 2009).

Goodreads.com. "Quotes by Ronald Reagan." http://www.goodreads.com/author/quotes/3543.Ronald_Reagan (accessed June 25, 2009).

Gray, Peter. "Reasons to Consider a Less Selective, Less Expensive College: Saving Money is Just One of Them." Psychology Today, 28 June 2009. http://www.psychologytoday.com/blog/freedom-learn/200810/reasons-consider-less-selective-less-expensive-college-saving-money-is-jus (accessed July 1, 2009).

Gula, Robert J. *Nonsense: A Handbook of Logical Fallacies*. New York: Axios Press, 2002.

Harwood, Matthew. "Murders at Schools Drop." Security Management Homepage. http://www.securitymanagement.com/news/murders-schools-drop (accessed June 2, 2009).

"Heart Disease Statistics." MamasHealth.com. http://www.mamashealth.com/HEART_stat.asp (accessed June 2, 2009).

Herszenhorn, David M. "Estimates of Iraq War Cost Were Not Close to Ballpark." New York Times Advertisement, 19 March 2008. http://www.nytimes.com/2008/03/19/washington/19cost.html (accessed June 3, 2009).

Hill, Austin, "Prop. 105: Good Intent, Bad Results." EastValleyTribune.com. http://www.eastvalleytribune.com/story/127667 (accessed June 4, 2009).

Hoerr, Thomas R. *Becoming a Multiple Intelligences School.* Alexandria, VA: Association For Supervision & Curriculum Development, 2000.

Holocaust Educational Resource. "Fallacy: Ad Hominem Tu Quoque." http://www.nizkor.org/features/fallacies/ad-hominem-tu-quoque.html (accessed June 3, 2009).

Holocaust Educational Resource. "Fallacy: Ad Hominem." http://www.nizkor.org/features/fallacies/ad-hominem.html (accessed June 3, 2009).

Holther, W. Ward, and W. B. Fearnside. *Fallacy: The Counterfeit of Argument*. New York: Spectrum, 1959.

Hornick, Ed, and Alexander Marquardt. "Obama Says Offshore Drilling Stance Nothing New." CNN.com. http://politicalticker.blogs.cnn.com/2008/08/02/obama-defends-offshore-oil-drilling-policy-change/ (accessed June 24, 2009).

The Huffington Post. "McCain On Obama: Clear Who Hamas Wants to be the Next President." http://www.huffingtonpost.com/2008/04/25/mccain-on-obama-clear-who_n_98614.html (accessed June 24, 2009).

Hurley, Patrick J. *A Concise Introduction to Logic*. Belmont, CA: Thomson Wadsworth, 2007.

Inside Higher Ed. "SAT Math Scores Are Up." http://www.insidehighered.com/news/2005/08/31/sat (accessed June 2, 2009).

Internet Encyclopedia of Philosophy (IEP), s.v., "Fallacies." http://www.iep.utm.edu/f/fallacy.htm#Common%20Cause (accessed June 13, 2009).

James Randi Educational Foundation. "Challenge Applications." JREF Forum. http://forums.randi.org/forumdisplay.php?f=43&sort=lastpost&order=desc&page=6 (accessed June 13, 2009).

James Randi Educational Foundation. "Challenge Info." http://www.randi.org/site/index.php/1m-challenge.html (accessed June 3, 2009).

James Randi Educational Foundation. "Million Dollar Challenge FAQ." http://www.randi.org/site/index.php/1m-challenge/challenge-faq.html (accessed June 3, 2009).

Johnston, Lauren. "CIA Faulted For Iraq Flaws." CBS News. http://www.cbsnews.com/stories/2004/07/09/iraq/main628550.shtml (accessed June 4, 2009).

Jokes Duniya. "Lipstick prints on mirror." http://www.jokesduniya.com/3916/lipstick-prints-on-mirror/ (accessed June 4, 2009).

Kahane, Howard. *Logic and Contemporary Rhetoric: The Use of Reason in Everyday Life*. Belmont, CA: Wadsworth, 1976.

LegalDaily.com. "U.S. Legal Memorandum on Torture of Terror Suspects and Detainees." http://www.legaldaily.com/ (accessed June 13, 2009).

Leiby, Richard. "A Couch Tom Cruise Won't Jump On." Washingtonpost.com. http://www.washingtonpost.com/wp-dyn/content/article/2005/06/24/AR2005062401866.html (accessed June 3, 2009).

Logical Fallacies. "A Post Hoc Fallacy." http://www.logicalfallacies.info/presumption/post-hoc/ (accessed June 2, 2009).

Logical Fallacies. "The Texas Sharpshooter Fallacy." The Fallacy Files. http://www.fallacyfiles.org/texsharp.html (accessed June 3, 2009).

Los Angeles Times. "Waterboarding is Legal, White House Says." http://www.latimes.com/news/nationworld/nation/la-na-torture7feb07,1,3156438.story (accessed June 4, 2009).

"McCain: Not Offended By Alleged Age-Baiting: Political Radar." ABC News. http://blogs.abcnews.com/politicalradar/2008/05/mccain-not-offe.html (accessed June 24, 2009).

McCoy, Alfred W. "A Question of Torture: CIA." Google Book Search, http://books.google.com/books?id=-sa-nUI3HGMC&pg=PA113&lpg=PA113&dq=Bush:+%22illegal+enemy+combat-ants%22&source=bl&ots (accessed June 4, 2009).

Media Partners. "Groupthink video." http://www.media-partners.com/communication_skills/groupthink.htm (accessed June 4, 2009).

Michigan.gov. "Pop Density-Land Area 24 Cities." http://74.125.95.132/search?q=cache:TUuwlc0Xb3YJ:www.michigan.gov/documents/hal_lm_census_landarea_ (accessed June 25, 2009).

Montecito Journal. "Ward Connerly." http://www.montecitojournal.net/archive/13/7/758/ (accessed June 3, 2009).

MSN Encarta: Online Dictionary, s.v. "logic." http://encarta.msn.com/dictionary_1861699377/logic.html (accessed June 2, 2009).

MSN.com. "Bush: U.S.A. Does Not Torture People." http://www.msnbc.msn.com/id/21148801/ (accessed June 4, 2009).

MSN.com. "Clinton offers universal health care plan." http://www.msnbc.msn.com/id/20819827/ (accessed June 3, 2009).

Myers, David G. "*Intuition: Its Powers and Perils (Yale Nota Bene).* New Haven: Yale University Press, 2004.

Myers, David. *Social Psychology*, 8th ed. New York: McGraw-Hill, 2005.

Nine, Gary. "Respect for Regular Folks: Arizona K–12 Arizona Education Points." http://www.respectforregularfolks.com/index.php/home/34-content/115-arizona-k-12-arizona-education-points (accessed June 25, 2009).

Office-humour.co.uk. "English professor writes..." http://office-humour.co.uk/item/6312/ (accessed June 24, 2009).

Overbye, Dennis. "Philosophers Notwithstanding, Kansas School Board Redefines Science." The New York Times, 15 November 2005. http://www.nytimes.com/2005/11/15/science/sciencespecial2/15evol.html?pagewanted=print (accessed June 4, 2009).

Paranormality.com. "Barnum Effect." http://www.paranormality.com/barnum_effect.shtml (accessed June 3, 2009).

Pearson, Jeannette Walls. "Psychiatrists Think Cruise Should Have Head Examined." The Ross Institute Internet Archives for the Study of Destructive Cults. http://www.rickross.com/reference/scientology/celebrities/celebrities39.html (accessed June 4, 2009).

Pirie, Madsen. *How to Win Every Argument: The Use and Abuse of Logic.* New York: Continuum International Publishing Group, 2007.

Quotatio. "Oliver Wendell Holmes Quotations: Index 4." http://www.quotatio.com/h/holmes-oliver-wendell-quotes.html (accessed June 4, 2009).

Quotations for Creative Thinking. "Creative Quotations from Burton Hillis." http://creativequotations.com/one/2313.htm (accessed June 2, 2009).

The Quotations Page. "Adolf Hitler: What luck for rulers..." http://www.quotationspage.com/quote/971.html (accessed June 2, 2009).

The Quotations Page. "Moses Hadas Quotes." http://www.quotationspage.com/quotes/Moses_Hadas/ (accessed June 24, 2009).

The Quotations Page. " Quote Details: Father Larry Lorenzoni: The Average Person Thinks." http://www.quotationspage.com/quote/993.html (accessed June 4, 2009).

RinkWorks. "Things People Said: Car Accident Reports." http://www.rinkworks.com/said/insurance.shtml (accessed June 24, 2009).

Robb, Robert. "Vote NO on Prop 105." http://thevotersofaz.com/?cat=4 (accessed June 24, 2009).

Roberts, C. K. "Weight Loss with a Low-Carbohydrate, Mediterranean, or Low-Fat Diet." The New England Journal of Medicine. http://content.nejm.org/cgi/content/full/359/3/229 (accessed June 2, 2009).

Sanchez, Claudio. "SAT Math Scores Peak; Verbal Remains Flat." National Public Radio (NPR). http://www.npr.org/templates/story/story.php?storyId=4824559 (accessed June 2, 2009).

Schwinn, Steven D., Ruthann Robson, and Nareissa L. Smith. "White House to Release Details of Enhanced Interrogation Techniques." USLaw.com. http://www.uslaw.com/library/Constitutional_Law/White_House_Release_Details_Enhanced_Interrogation_Techniques.php?item=422110 (accessed June 4, 2009).

Shaw, Mary. "Bush Admits He Would Have Gone to War Knowing What We Know Now." Associated Content. http://www.associatedcontent.com/article/17140/bush_admits_he_would_have_gone_to_war.html (accessed June 24, 2009).

The Skeptic's Dictionary. "Texas sharpshooter fallacy." http://skepdic.com/texas.html (accessed June 3, 2009).

Skier, Stephanie M. "Polaroid Corporation Files for Bankruptcy." The Harvard Crimson. http://www.thecrimson.com/article.aspx?ref=121699 (accessed June 24, 2009).

Smith, Tim. "Millions of superstitious readers-- and many athletes-- believe that an appearance on Sports Illustrated's cover is the kiss of death. But is there really such a thing as the SI Jinx? Sports Illustrated, 27 August 2003. http://sportsillustrated.cnn.com/2003/magazine/08/27/jinx/ (accessed June 3, 2009).

"Snopes.com. "Welfare Application Humor." http://www.snopes.com/humor/lists/welfare.asp (accessed June 24, 2009).

Sports Illustrated. "Alex Wolff: Unraveling the SI cover jinx," 29 August 2003. http://sportsillustrated.cnn.com/2003/writers/alexander_wolff/08/27/viewpoint_flashback/index.html (accessed June 3, 2009).

Surgery.org. "Facelift: Does Looking Younger Help You Live Longer?" http://www.surgery.org/press/news-release.php?iid=168 (accessed June 2, 2009).

Sympatico. "Paradigms." http://www3.sympatico.ca/saburns/pg0203.htm (accessed June 2, 2009).

Taylor, Shelly E. "Positive Illusions: Creative Self-Deception and the Healthy Mind." http://www.amazon.com/Positive-Illusions-Creative-Self-Deception-Healthy/dp/0465060536 (accessed June 25, 2009).

Teeple, Amy. "BMI Calculator , Body Mass Index Chart , Ideal Body Weight Chart Review." DocShop.com. http://www.docshop.com/2008/04/08/arnold-schwarzenegger-is-obese-problems-with-body-mass-index-bmi-calculations/ (accessed June 2, 2009).

"Texas Bills Would Allow Guns at College." Msnbc.com (Education). http://www.msnbc.msn.com/id/29944438/ (accessed June 2, 2009).

ThinkExist.com Quotations. "Mark Twain Quotes." http://thinkexist.com/quotation/all_generalizations_are_false-including_this_one/216596.html (accessed June 2, 2009).

ThinkExist.com Quotations. "Ayn Rand quotes." http://thinkexist.com/quotation/wealth_is_the_product_of_man-s_capacity_to_think/220796.html (accessed June 2, 2009).

ThinkExist.com. "Brandi Snyder Quotes." http://thinkexist.com/quotation/to_the_world_you_may_be_just_one_person-but_to/189999.html (accessed June 25, 2009).

ThinkExist.com. "Groucho Marx quotes." http://thinkexist.com/quotation/i_refuse_to_join_any_club_that_would_have_me_as_a/207365.html (accessed June 24, 2009).

ThinkExist.com. "Mark Twain quotes." http://thinkexist.com/quotation/the_difference_between_the_right_word_and_the/11504.html (accessed June 4, 2009).

ThinkQuest Library. "Tiger Woods." http://library.thinkquest.org/5672/Tiger_Woods.html (accessed June 4, 2009).

Three Trillion Dollar War: The True Cost of the Iraq Conflict, http://threetrilliondollarwar.org/ (accessed June 3, 2009).

Time Magazine. "Can't Make Up Your Mind?" 20 October 2008.

Time Magazine. "What Kind of Voter Are You?" 20 October 2008.

Truman Library. "Clark M. Clifford Oral History Interview, February 14, 1973." http://www.trumanlibrary.org/oralhist/cliford7.htm (accessed June 24, 2009).

U.S. Census Bureau. "State Ranking--Statistical Abstract of the United States." www.census.gov/statab/ranks/rank01.html (accessed June 24, 2009).

United States Department of Justice. "Legal Standards Applicable Under 18 U.S.C. §§ 2340-2340A." http://www.usdoj.gov/olc/18us (accessed June 28, 2009).

US Census Bureau. "US Census Press Releases." Census Bureau Home Page. http://www.census.gov/Press-Release/www/releases/archives/facts_for_features_special_editions/007108.html (accessed June 2, 2009).

US News & World Report. "How Much is That College Degree Really Worth?" 30 October 2008. http://www.usnews.com/articles/education/2008/10/30/how-much-is-that-college-degree-really-worth.html (accessed June 3, 2009).

USATODAY.com. "Statements on Atkins' death." http://www.usatoday.com/news/health/2004-02-10-atkins-statements_x.htm (accessed June 3, 2009).

Vaughan, Susan C., M.D. *Half Empty Half Full: Understanding the Psychological Roots of Optimism.* New York: Harcourt, 2000.

Walton, Douglas N. *Plausible Argument in Everyday Conversation* (S U N Y Series in Speech Communication). Albany, NY: State University of New York Press, 1992.

Wang, Penelope. "The real return on your college investment." CNNMoney.com. http://money.cnn.com/2008/08/20/pf/college/college_price.moneymag/index3.htm (accessed June 3, 2009).

WebMD. "Drink More Diet Soda, Gain More Weight?" http://www.webmd.com/diet/news/20050613/drink-more-diet-soda-gain-more-weight (accessed June 2, 2009).

Welna, David. "McCain and Bush: Common Ground on War, Torture." National Public Radio (NPR). http://www.npr.org/templates/story/story.php?storyId=91560330 (accessed June 3, 2009).

Wikipedia, s.v. "list of cognitive biases." http://en.wikipedia.org/wiki/List_of_cognitive_biases (accessed June 4, 2009).

www.ingramcontent.com/pod-product-compliance
Lightning Source LLC
LaVergne TN
LVHW080033210926
848481LV00011B/3171
* 9 7 8 0 7 5 7 5 7 1 0 1 5 *